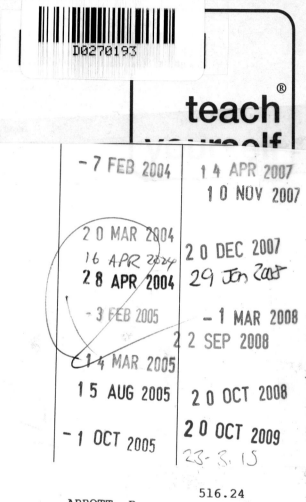

teach®

ABBOTT, P.
Trigonometry

teach yourself®

trigonometry
p. abbott
revised by hugh neill

For over 60 years, more than 40 million people have learnt over 750 subjects the **teach yourself** way, with impressive results.

be where you want to be
with **teach yourself**

For UK order enquiries: please contact Bookpoint Ltd., 130 Milton Park, Abingdon, Oxon OX14 4SB. Telephone: +44 (0)/1235 827720. Fax: +44 (0)/1235 400454. Lines are open 09.00–18.00, Monday to Saturday, with a 24-hour message answering service. You can also order through our website www.madaboutbooks.com.

For USA order enquiries: please contact McGraw-Hill Customer Services, PO Box 545, Blacklick, OH 43004-0545, USA. Telephone: 1-800-722-4726. Fax: 1-614-755-5645.

For Canada order enquiries: please contact McGraw-Hill Ryerson Ltd., 300 Water St, Whitby, Ontario L1N 9B6, Canada. Telephone: 905 430 5000. Fax: 905 430 5020.

Long renowned as the authoritative source for self-guided learning – with more than 30 million copies sold worldwide – the Teach Yourself series includes over 300 titles in the fields of languages, crafts, hobbies, business, computing and education.

British Library Cataloguing in Publication Data: A catalogue record for this title is available from The British Library

Library of Congress Catalog Card Number: On file

First published in UK 1992 by Hodder Headline Ltd., 338 Euston Road, London, NW1 3BH.

First published in US 1992 by Contemporary Books, A Division of The McGraw-Hill Companies, 1 Prudential Plaza, 130 East Randolph Street, Chicago, IL 60601 USA.

Typeset by Pantek Arts Ltd, Maidstone, Kent.

Printed in Great Britain for Hodder & Stoughton Educational, a division of Hodder Headline Ltd., 338 Euston Road, London NW1 3BH by Cox & Wyman Ltd., Reading, Berkshire.

Impression number   10 9 8 7 6 5 4 3 2 1

Year   2009 2008 2007 2006 2005 2004 2003

contents

preface

Teach Yourself Trigonometry has been substantially revised and rewritten to take account of modern needs and recent developments in the subject.

It is anticipated that every reader will have access to a scientific calculator which has sines, cosines and tangents, and their inverses. It is also important that the calculator has a memory, so that intermediate results can be stored accurately. No support has been given about how to use the calculator, except in the most general terms. Calculators vary considerably in the keystrokes which they use, and what is appropriate for one calculator may be inappropriate for another.

There are many worked examples in the book, with complete, detailed answers to all the questions. At the end of each worked example, you will find the symbol ∎ to indicate that the example has been completed, and what follows is text.

Some of the exercises from the original Teach Yourself Trigonometry have been used in this revised text, but all the answers have been reworked to take account of the greater accuracy available with calculators.

I would like to thank Linda Moore for her help in reading and correcting the text. But the responsibility for errors is mine.

Hugh Neill
November 1997

01

historical
background

In this chapter you will learn:
- what trigonometry is
- a little about its origins.

1.1 Introduction

One of the earliest known examples of the practical application of geometry was the problem of finding the height of one of the Egyptian pyramids. This was solved by Thales, the Greek philosopher and mathematician (*c.* 640 BC to 550 BC) using similar triangles.

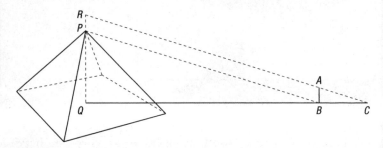

figure 1.1

Thales observed the length of the shadows of the pyramid and a stick, *AB*, placed vertically into the ground at the end of the shadow of the pyramid, shown in Figure 1.1.

QB and *BC* represent the lengths of the shadow of the pyramid and the stick. Thales said 'The height of the pyramid is to the length of the stick, as the length of the shadow of the pyramid is to the length of the shadow of the stick.'

That is, in Figure 1.1,

$$\frac{PQ}{AB} = \frac{QB}{BC}.$$

As *QB*, *AB*, and *BC* are known, you can calculate *PQ*.

We are told that the king, Amasis, was amazed at Thales' application of an abstract geometrical principle to the solution of such a problem.

This idea is taken up in Chapter 02, in introducing the idea of the tangent ratio.

1.2 What is trigonometry?

Trigonometry is the branch of mathematics which deals with the relationships between sides and angles of triangles.

For example, if you have a right-angled triangle with sides 7 cm and 5 cm as shown in Figure 1.2, trigonometry enables you to calculate the angles of the triangle and the length of the other side.

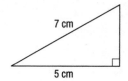

7 cm

5 cm

figure 1.2

You should be able to do these calculations when you have studied Chapter 02.

Similarly, if you know the information given in Figure 1.3 about the triangle *ABC*, you can calculate the area of the triangle, the length of the other side, and the magnitude of the other angles.

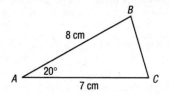

B

8 cm

A 20°

7 cm

C

figure 1.3

You should be able to do this when you have studied Chapter 07.

1.3 The origins of trigonometry

According to George Gheverghese Joseph in his book, *The Crest of the Peacock*, published by Penguin Books in 1990, the origins of trigonometry are obscure. A systematic study of the relationships between the angle at the centre of a circle and the length of chord subtending it seems to have started with Hipparchus (*c*. 150 BC). See Figure 1.4.

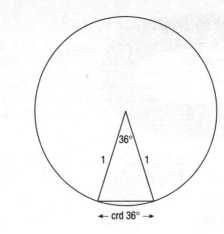

figure 1.4

Tables of the lengths of chords for given angles were produced by Ptolemy (*c*. AD 100).

Trigonometry began to resemble its present form from about AD 150 with the work of Aryabhata I. The knowledge was transmitted to the Arabs and thence to Europe, where a detailed account of trigonometric knowledge appeared under the title *De triangulis omni modis*, written in 1464 by Regiomontanus.

Originally trigonometry may have been used to measure the steepness of the faces of the pyramids, but it was also used for astronomy.

The Indian mathematicians Varahamihara (*c*. AD 500) and Aryabhata I published tables from which it is possible to compute values of sines of angles. These values are extremely accurate. For example, their values for sin 45° were respectively 0.70708 and 0.70710, compared with the modern value, 0.70711.

the tangent

In this chapter you will learn:
- what a tangent is
- the meanings of 'opposite', 'adjacent' and 'hypotenuse' in right-angled triangles
- how to solve problems using tangents.

2.1 Introduction

The method used by Thales to find the height of the pyramid in ancient times is essentially the same as the method used today. It is therefore worth examining more closely. Figure 2.1 is the same as Figure 1.1.

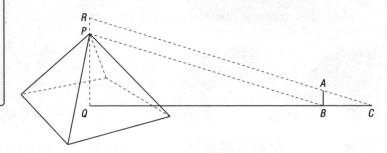

figure 2.1

You can assume that the sun's rays are parallel because the sun is a long way from the earth. In Figure 2.1 it follows that the lines RC and PB which represent the rays falling on the tops of the objects are parallel.

Therefore, angle PBQ = angle ACB (they are corresponding angles). These angles each represent the altitude of the sun.

As angles PQB and ABC are right angles, triangles PQB and ABC are similar, so

$$\frac{PQ}{QB} = \frac{AB}{BC} \text{ or } \frac{PQ}{AB} = \frac{QB}{BC}.$$

The height PQ of the pyramid is independent of the length of the stick AB. If you change the length AB of the stick, the length of its shadow will be changed in proportion. You can therefore make the following important general deduction.

For the given angle ACB, the ratio $\dfrac{AB}{BC}$ stays constant whatever the length of AB. You can calculate this ratio beforehand for any angle ACB. If you do this, you do not need to use the stick, because if you know the angle and the value of the ratio, and you have measured the length QB, you can calculate PQ.

Thus if the angle of elevation is 64° and the value of the ratio for this angle had been previously found to be 2.05, then you have

$$\frac{PQ}{QB} = 2.05.$$

Therefore $\qquad PQ = 2.05 \times QB.$

2.2 The idea of the tangent ratio

The idea of a constant ratio for every angle is the key to the development of trigonometry.

Let POQ (Figure 2.2) be any acute angle $\theta°$. From points A, B, C on one arm, say OQ, draw perpendiculars AD, BE, CF to the other arm, OP. As these perpendiculars are parallel, the triangles AOD, BOE and COF are similar.

Therefore

$$\frac{AD}{OD} = \frac{BE}{OE} = \frac{CF}{OF}.$$

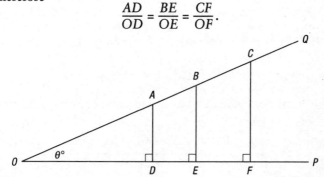

figure 2.2

Now take any point Y, it does not matter which, on the arm OQ. For that angle $\theta°$ the ratio of the perpendicular XY drawn from Y on the arm OQ to the distance OX intercepted on the other arm OP is constant. See Figure 2.3.

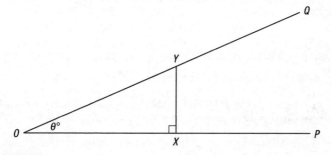

figure 2.3

This is true for any angle; each angle $\theta°$ has its own particular ratio corresponding to it. This ratio is called the **tangent** of the angle $\theta°$.

In practice, the name tangent is abbreviated to **tan**.

Thus for $\theta°$ in Figures 2.2 and 2.3 you can write

$$\tan \theta° = \frac{XY}{OX}.$$

2.3 A definition of tangent

There was a general discussion of the idea of the tangent ratio in Section 2.2, but it is important to refine that discussion into a formal definition of the tangent of an angle.

In Figure 2.4, the origin O is the centre of a circle of radius 1 unit. Draw a radius OP at an angle $\theta°$ to the x-axis, where $0 \leqslant \theta < 90$. Let the coordinates of P be (x, y).

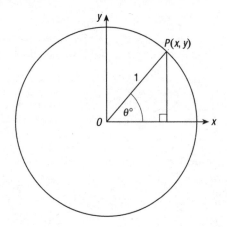

figure 2.4

Then the **tangent** of the angle $\theta°$, written $\tan \theta°$, is defined by

$$\tan \theta° = \frac{y}{x}.$$

You can see from the definition that if $\theta = 0$ the y-coordinate of P is 0, so $\tan 0 = 0$.

If $\theta = 45$, then $x = y$, so

$$\tan 45° = 1.$$

As θ increases, y increases and x decreases, so the tangents of angles close to 90° are very large. You will see that when $\theta = 90$, the value of x is 0, so $\frac{y}{x}$ is not defined; it follows that $\tan 90°$ does not exist, and is undefined.

2.4 Values of the tangent

You can find the value of the tangent of an angle by using your calculator. Try using it. You should find that the tangent of 45°, written $\tan 45°$, is 1, and $\tan 60° = 1.732...$. If you have difficulty with this you should consult your calculator handbook, and make sure that you can find the tangent of any angle quickly and easily.

Your calculator must be in the correct mode. There are other units, notably radians or rads, for measuring angle, and you must ensure that your calculator is in degree mode, rather than radian or rad mode. Radians are widely used in calculus, and are the subject of Chapter 05.

Some calculators also give tangents for grades, another unit for angle. There are 100 grades in a right angle; this book will not use grades.

Your calculator will also reverse this process of finding the tangent of an angle. If you need to know which angle has a tangent of 0.9, you look up the **inverse tangent**. This is often written as $\tan^{-1} 0.9$, or sometimes as $\arctan 0.9$. Check that $\tan^{-1} 0.9 = 41.987...°$. If it is not, consult your calculator handbook.

In the work that follows, the degree sign will always be included, but you might wish to leave it out in your work, provided there is no ambiguity. Thus you would write $\tan 45° = 1$ and $\tan 60° = 1.732...$.

Exercise 2.1

In questions 1 to 6, use your calculator to find the values of the tangents of the following angles. Give your answers correct to three decimal places.

1	$\tan 20°$	**2**	$\tan 30°$
3	$\tan 89.99°$	**4**	$\tan 40.43°$
5	$\tan 62°$	**6**	$\tan 0.5°$

In questions 7 to 12, use your calculator to find the angles with the following tangents. Give your answer correct to the nearest one hundredth of a degree.

7	0.342	**8**	2
9	6.123	**10**	0.0001
11	1	**12**	$\sqrt{3}$

2.5 Notation for angles and sides

Using notation such as ABC for an angle is cumbersome. It is often more convenient to refer to an angle by using only the middle letter of the three which define it. Thus, if there is no ambiguity, $\tan B$ will be used in preference for $\tan ABC$.

Single Greek letters such as α (alpha), β (beta), θ (theta) and ϕ (phi) are often used for angles.

Similarly, it is usually easier to use a single letter such as h to represent a distance along a line, rather than to give the beginning and end of the line as in the form AB.

2.6 Using tangents

Here are some examples which illustrate the use of tangents and the technique of solving problems with them.

Example 2.1

A surveyor who is standing at a point 168 m horizontally distant from the foot of a tall tower measures the angle of elevation of the top of the tower as 38.25°. Find the height above the ground of the top of the tower.

You should always draw a figure. In Figure 2.5, P is the top of the tower and Q is the bottom. The surveyor is standing at O which is at the same level as Q. Let the height of the tower be h metres.

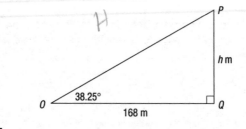

figure 2.5

Then angle POQ is the angle of elevation and equals 38.25°.

Then $$\frac{h}{168} = \tan 38.25°$$

$$h = 168 \times \tan 38.25°$$

$$= 168 \times 0.7883364...$$

$$= 132.44052... .$$

The height of the tower is 132 m, correct to three significant figures. ∎

In practice, if you are using a calculator, there is no need to write down all the steps given above. You should write down enough so that you can follow your own working, but you do not need to write down the value of the tangent as an intermediate step. It is entirely enough, and actually better practice, to write the calculation above as

$$\frac{h}{168} = \tan 38.25°$$

$$h = 168 \times \tan 38.25°$$

$$= 132.44052... .$$

However, in this chapter and the next, the extra line will be inserted as a help to the reader.

Example 2.2

A person who is 168 cm tall had a shadow which was 154 cm long. Find the angle of elevation of the sun.

In Figure 2.6 let PQ be the person and OQ be the shadow. Then PO is the sun's ray and $\theta°$ is the angle of elevation of the sun.

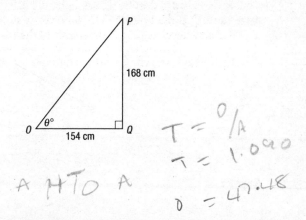

figure 2.6

Then
$$\tan \theta° = \frac{168}{154}$$

$$= 1.09090\ldots$$

$$\theta = \tan^{-1} 1.09090\ldots$$

$$= 47.489\ldots .$$

Therefore the angle of elevation of the sun is approximately 47.49°. ∎

Note once again that you can use the calculator and leave out a number of steps, provided that you give enough explanation to show how you obtain your result. Thus you could write

$$\tan \theta° = \frac{168}{154}$$

$$\theta = \tan^{-1}\left(\frac{168}{154}\right)$$

$$= 47.489\ldots .$$

Example 2.3
Figure 2.7 represents a section of a symmetrical roof in which *AB* is the span, and *OP* the rise. *P* is the mid-point of *AB*.

The rise of the roof is 7 m and its angle of slope is 32°. Find the roof span.

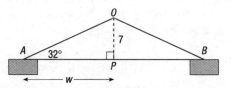

figure 2.7

As the roof is symmetrical, *OAB* is an isosceles triangle, so *OP* is perpendicular to *AB*. Call the length *AP*, *w* metres.

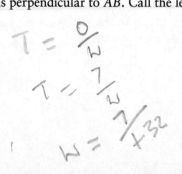

Therefore: $\tan 32° = \dfrac{7}{w}$,

so
$$w = \dfrac{7}{\tan 32°}$$

$$= \dfrac{7}{0.624869\ldots}$$

$$= 11.2023\ldots .$$

The roof span is $2w$ metres, that is approximately 22.4 m. ∎

Exercise 2.2

1 The angle of elevation of the sun is 48.4°. Find the height of a flag staff whose shadow is 7.42 m long.
2 A boat leaving a harbour travels 4 miles east and 5 miles north. Find the bearing of the boat from the harbour.
3 A boat which is on a bearing of 038° from a harbour is 6 miles north of the harbour. How far east is the boat from the harbour?
4 A ladder resting against a wall makes an angle of 69° with the ground. The foot of the ladder is 7.5 m from the wall. Find the height of the top of the ladder.
5 From the top window of a house which is 1.5 km away from a tower it is observed that the angle of elevation of the top of the tower is 3.6° and the angle of depression of the bottom is 1.2°. Find the height of the tower in metres.
6 From the top of a cliff 32 m high it is noted that the angles of depression of two boats lying in the line due east of the cliff are 21° and 17°. How far are the boats apart?
7 Two adjacent sides of a rectangle are 15.8 cm and 11.9 cm. Find the angles which a diagonal of the rectangle makes with the sides.
8 P and Q are two points directly opposite to one another on the banks of a river. A distance of 80 m is measured along one bank at right angles to PQ. From the end of this line the angle subtended by PQ is 61°. Find the width of the river.
9 A ladder which is leaning against a wall makes an angle of 70° with the ground and reaches 5 m up the wall. The foot of the ladder is then moved 50 cm closer to the wall. Find the new angle that the ladder makes with the ground.

2.7 Opposite and adjacent sides

Sometimes the triangle with which you have to work is not conveniently situated on the page. Figure 2.8a shows an example of this.

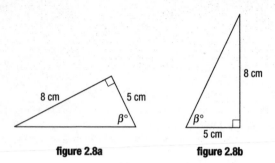

figure 2.8a　　　　**figure 2.8b**

In this case, there is no convenient pair of axes involved. However, you could rotate the figure, either actually or in your imagination, to obtain Figure 2.8b.

You can now see that $\tan\beta° = \frac{8}{5} = 1.6$, and you can calculate β, but how could you see that $\tan\beta° = \frac{8}{5} = 1.6$ easily from the diagram in Figure 2.8a, without going through the process of getting to Figure 2.8b?

When you are using a right-angled triangle you will always be interested in one of the angles other than the right angle. For the moment, call this angle the 'angle of focus'. One of the sides will be opposite this angle; call this side the **opposite**. One of the other sides will join the angle in which you are interested; call this side the **adjacent**.

Then
$$\text{tangent} = \frac{\text{opposite}}{\text{adjacent}}.$$

This works for all right-angled triangles. In the two cases in Figure 2.8a and 2.8b, the opposite and adjacent sides are labelled in Figure 2.9a and 2.9b.

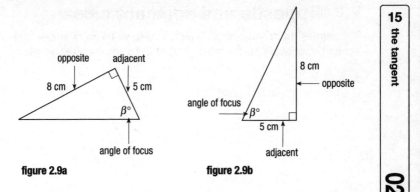

figure 2.9a   **figure 2.9b**

As you can see, in both cases

$$\tan\beta° = \frac{\text{opposite}}{\text{adjacent}} = \frac{8}{5}.$$

Many people find this method the most convenient when using the tangent.

The other side of the right-angled triangle, the longest side, is called the **hypotenuse**. The hypotenuse will feature in Chapter 03.

Example 2.4

In a triangle ABC, angle $B = 90°$, $AB = 5$ cm and $BC = 7$ cm. Find the size of angle A.

Draw a diagram, Figure 2.10.

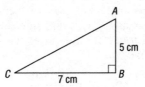

figure 2.10

In triangle ABC, focus on angle A. The opposite is 7 cm and the adjacent is 5 cm.

Therefore $\tan A° = \frac{7}{5}$

and   angle A = 54.46. ▮

Note that in this case you could find angle C first using $\tan C° = \frac{5}{7}$, and then use the fact that the sum of the angles of triangle ABC is $180°$ to find angle A.

Example 2.5
Find the length a in Figure 2.11.

$T = \frac{O}{A}$

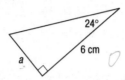

figure 2.11

Focus on the angle $24°$. The opposite side is a and the adjacent side is $6\,cm$.

Then
$$\tan 24° = \frac{a}{6}$$
$$a = 6 \tan 24°$$
$$= 6 \times 0.4452\ldots$$
$$= 2.671\ldots .$$

Therefore $a = 2.67\,cm$, correct to 3 significant figures. ∎

Example 2.6
Find the length x cm in Figure 2.12.

$T = \frac{O}{A}$

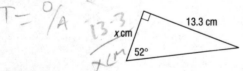

figure 2.12

Focus on the angle $52°$. The opposite side is $13.3\,cm$ and the adjacent side is $x\,cm$.

$$\tan 52° = \frac{13.3}{x}$$

Then
$$\tan 52° = \frac{13.3}{x}$$

$$x = \frac{13.3}{\tan 52°}$$

$$= \frac{13.3}{1.2799\ldots}$$

$$= 10.391\ldots .$$

Therefore the length is 10.4 cm, correct to 3 significant figures. ∎

Exercise 2.3
In each of the following questions find the marked angle or side.

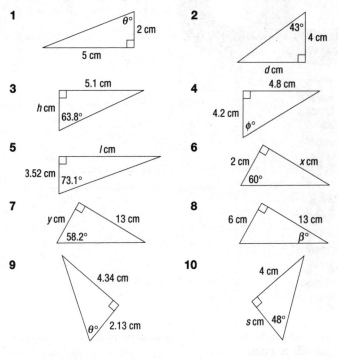

1 θ°, 2 cm, 5 cm

2 43°, 4 cm, d cm

3 5.1 cm, h cm, 63.8°

4 4.8 cm, 4.2 cm, φ°

5 l cm, 3.52 cm, 73.1°

6 2 cm, x cm, 60°

7 y cm, 13 cm, 58.2°

8 6 cm, 13 cm, β°

9 4.34 cm, 2.13 cm, θ°

10 4 cm, s cm, 48°

03

sine and cosine

In this chapter you will learn:
- what the sine and cosine are
- how to use the sine and cosine to find lengths and angles in right-angled triangles
- how to solve multistage problems using sines, cosines and tangents.

3.1 Introduction

In Figure 3.1 a perpendicular is drawn from A to OB.

You saw on page 8 that the ratio $\dfrac{AB}{OB} = \tan\theta°$.

Now consider the ratios of each of the lines AB and OB to the hypotenuse OA of triangle OAB.

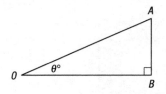

figure 3.1

Just as for a fixed angle $\theta°$ the ratio $\dfrac{AB}{OB}$ is constant, (and equal to $\tan\theta°$) wherever A is, so also the ratio $\dfrac{AB}{OA}$, that is $\dfrac{\text{opposite}}{\text{hypotenuse}}$, is constant.

This ratio is called the **sine of the angle** $\theta°$ and is written $\sin\theta°$.

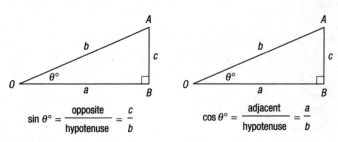

$$\sin\theta° = \frac{\text{opposite}}{\text{hypotenuse}} = \frac{c}{b} \qquad \cos\theta° = \frac{\text{adjacent}}{\text{hypotenuse}} = \frac{a}{b}$$

figure 3.2

Similarly, the ratio $\dfrac{OB}{OA}$, that is $\dfrac{\text{adjacent}}{\text{hypotenuse}}$, is also constant for the angle $\theta°$.

This ratio is called the **cosine of the angle** $\theta°$ and is written $\cos\theta°$.

Thus $\quad \sin\theta° = \dfrac{\text{opposite}}{\text{hypotenuse}} = \dfrac{c}{b}, \quad \cos\theta° = \dfrac{\text{adjacent}}{\text{hypotenuse}} = \dfrac{a}{b}.$

3.2 Definition of sine and cosine

In Section 3.1 there is a short discussion introducing the sine and cosine ratios. In this section there is a more formal definition.

Draw a circle with radius 1 unit, and centre at the origin O. Draw the radius OP at an angle $\theta°$ to the x-axis in an anti-clockwise direction. See Figure 3.3.

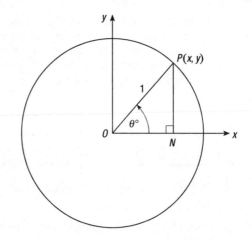

figure 3.3

Then let P have coordinates (x, y).

Then $\sin \theta° = y$ and $\cos \theta° = x$ are the definitions of sine and cosine which will be used in the remainder of the book.

Note the arrow labelling the angle $\theta°$ in Figure 3.3; this is to emphasize that angles are measured positively in the anti-clockwise direction.

Note also two other properties of $\sin \theta°$ and $\cos \theta°$.

- In the triangle OPN, angle $OPN = (90 - \theta)°$, and

$$\sin(90 - \theta)° = \frac{\text{opposite}}{\text{hypotenuse}} = \frac{x}{1} = \cos \theta°,$$

that is $\qquad \sin(90 - \theta)° = \cos \theta°.$

Similarly $\qquad \cos(90 - \theta)° = \sin \theta°.$

- Using Pythagoras's theorem on triangle *OPN* gives $x^2 + y^2 = 1$.

  Therefore $\sin^2 \theta° + \cos^2 \theta° = 1$,

  where $\sin^2 \theta°$ means $(\sin \theta°)^2$ and means $\cos^2 \theta°$ means $(\cos \theta°)^2$.

The equation $\sin^2 \theta° + \cos^2 \theta° = 1$ is often called Pythagoras's theorem.

Finding the values of the sine and cosine of angles is similar to finding the tangent of an angle. Use your calculator in the way that you would expect. You can use the functions $\sin^{-1}$ and $\cos^{-1}$ to find the inverse sine and cosine in the same way that you used $\tan^{-1}$ to find the inverse tangent.

3.3 Using the sine and cosine

In the examples which follow there is a consistent strategy for starting the problem.

- Look at the angle (other than the right angle) involved in the problem.

- Identify the sides, adjacent, opposite and hypotenuse, involved in the problem.

- Decide which trigonometric ratio is determined by the two sides.

- Make an equation which starts with the trigonometric ratio for the angle concerned, and finishes with the division of two lengths.

- Solve the equation to find what you need.

Here are some examples which use this strategy.

Example 3.1
Find the length marked x cm in the right-angled triangle in Figure 3.4.

figure 3.4

The angle concerned is 51°; relative to the angle of 51°, the side 2.5 cm is the adjacent, and the side marked x cm is the hypotenuse. The ratio concerned is the cosine.

Start by writing $\qquad \cos 51° = \dfrac{2.5}{x}$.

Then solve this equation for x.

$$x = \frac{2.5}{\cos 51°} = 3.972\ldots .$$

The length of the side is 3.97 cm approximately. ∎

Example 3.2

The length of each leg of a step ladder is 2.5 m. When the legs are opened out, the distance between their feet is 2 m. Find the angle between the legs.

In Figure 3.5, let AB and AC be the legs of the ladder. As there is no right angle, you have to make one by dropping the perpendicular AO from A to the base BC. The triangle ABC is isosceles, so AO bisects the angle BAC and the base BC. Therefore

$$BO = OC = 1\,\text{m}.$$

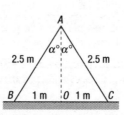

figure 3.5

You need to find angle BAC. Call it $2\alpha°$, so angle $BAO = \alpha°$.

The sides of length 1 m and 2.5 m are the opposite and the hypotenuse for the angle $\alpha°$, so you need the sine ratio.

Then $\qquad \sin \alpha° = \dfrac{1}{2.5} = 0.4$

so $\qquad\qquad \alpha = 23.578\ldots ,$

and $\qquad\qquad 2\alpha = 47.156\ldots .$

Therefore the angle between the legs is 47.16° approximately. ∎

Example 3.3

A 30 m ladder on a fire engine has to reach a window 26 m from the ground which is horizontal and level. What angle, to the nearest degree, must it make with the ground and how far from the building must it be placed?

Let the ladder be *AP* (Figure 3.6), let $\theta°$ be the angle that the ladder makes with the ground and let *d* metres be the distance of the foot of the ladder from the window.

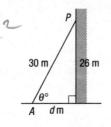

figure 3.6

As the sides 26 m and 30 m are the opposite and the hypotenuse for the angle $\theta°$, you need the sine ratio.

Then $$\sin \theta° = \frac{26}{30} = 0.8666\ldots$$

so $$\theta° = 60.073\ldots .$$

The ladder is placed at an angle of 60° to the ground.

To find the distance *d*, it is best to use Pythagoras's theorem.

$$d^2 = 30^2 - 26^2 = 900 - 676 = 244,$$

so $$d = 14.966\ldots .$$

The foot of the ladder is 14.97 m from the wall. ∎

Example 3.4

The height of a cone is 18 cm, and the angle at the vertex is 88°. Find the slant height.

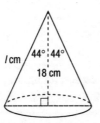

figure 3.7

In Figure 3.7, let l cm be the slant height of the cone. Since the perpendicular to the base bisects the vertical angle of the cone, each part is 44°.

The sides 18 cm and l cm are the adjacent and hypotenuse for the angle 44°, so the ratio concerned is the cosine.

Then
$$\cos 44° = \frac{18}{l},$$

so
$$l = \frac{18}{\cos 44°} = \frac{18}{0.71933\ldots}$$

$$l = 25.022\ldots .$$

The slant height is approximately 25.0 cm. ∎

Exercise 3.1

In questions 1 to 10, find the side or angle indicated by the letter.

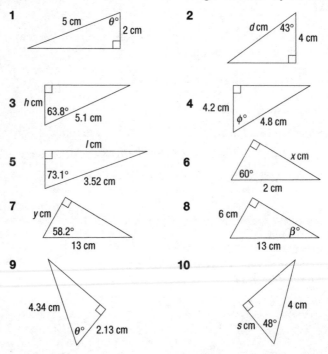

1 5 cm $\theta°$ 2 cm

2 d cm 43° 4 cm

3 h cm 63.8° 5.1 cm

4 4.2 cm $\phi°$ 4.8 cm

5 l cm 73.1° 3.52 cm

6 x cm 60° 2 cm

7 y cm 58.2° 13 cm

8 6 cm $\beta°$ 13 cm

9 4.34 cm $\theta°$ 2.13 cm

10 4 cm s cm 48°

11 A circle of radius 45 mm has a chord of length 60 mm. Find the sine and the cosine of the angle at the centre of the circle subtended by this chord.

12 In a circle with radius 4 cm, a chord is drawn subtending an angle of 80° at the centre. Find the length of this chord and its distance from the centre.

13 The sides of a triangle are 135 mm, 180 mm and 225 mm. Prove that the triangle is right-angled, and find its angles.

14 In a right-angled triangle, the hypotenuse has length 7.4 cm, and one of the other sides has length 4.6 cm. Find the smallest angle of the triangle.

15 A boat travels a distance of 14.2 km on a bearing of 041°. How far east has it travelled?

16 The height of an isosceles triangle is 3.8 cm, and the equal angles are 52°. Find the length of the equal sides.

17 A chord of a circle is 3 m long, and it subtends an angle of 63° at the centre of the circle. Find the radius of the circle.

18 A person is walking up a road angled at 8° to the horizontal. How far must the person walk along the road to rise a height of 1 km?

19 In a right-angled triangle the sides containing the right angle are 4.6 m and 5.8 m. Find the angles and the length of the hypotenuse.

3.4 Trigonometric ratios of 45°, 30° and 60°

You can calculate the trigonometric ratios exactly for some simple angles.

Sine, cosine and tangent of 45°

Figure 3.8 shows an isosceles right-angled triangle whose equal sides are each 1 unit.

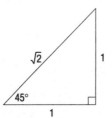

figure 3.8

Using Pythagoras's theorem, the hypotenuse has length $\sqrt{2}$ units.

Therefore the trigonometric ratios for 45° are given by

$$\sin 45° = \frac{1}{\sqrt{2}}, \cos 45° = \frac{1}{\sqrt{2}}, \tan 45° = 1.$$

You can, if you wish, use the equivalent values

$$\sin 45° = \frac{\sqrt{2}}{2}, \cos 45° = \frac{\sqrt{2}}{2}, \tan 45° = 1.$$

These values are obtained from their previous values by noting that

$$\frac{1}{\sqrt{2}} = \frac{1}{\sqrt{2}} \times \frac{\sqrt{2}}{\sqrt{2}} = \frac{\sqrt{2}}{2}.$$

Sine, cosine and tangent of 30° and 60°

Figure 3.9 shows an equilateral triangle of side 2 units.

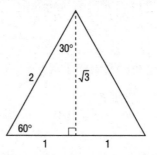

figure 3.9

The perpendicular from the vertex bisects the base, dividing the original triangle into two triangles with angles of 30°, 60° and 90° and sides of length 1 unit, 2 units and, using Pythagoras's theorem, $\sqrt{3}$ units.

Therefore the trigonometric ratios for 30° are given by

$$\sin 30° = \frac{1}{2}, \cos 30° = \frac{\sqrt{3}}{2}, \tan 30° = \frac{1}{\sqrt{3}} \text{ or } \frac{\sqrt{3}}{3}.$$

The same ratios for 60° are given by

$$\sin 60° = \frac{\sqrt{3}}{2}, \cos 60° = \frac{1}{2}, \tan 60° = \sqrt{3}.$$

It is useful either to remember these results, or to be able to get them quickly.

3.5 Using the calculator accurately

When you use a calculator to solve an equation such as $\sin \theta° = \frac{23}{37}$, it is important to be able to get as accurate an answer as you can.

Wrong method

$$\sin \theta° = \frac{23}{37} = 0.622$$

$$\sin \theta° = 38.46.$$

Correct method

$$\sin \theta° = \frac{23}{37} = 0.6216216216$$

$$\theta = 38.43.$$

What has happened? The problem is that in the wrong method the corrected answer, 0.622, a three-significant-figure approximation, has been used in the second part of the calculation to find the angle, and has introduced an error.

You can avoid the error by not writing down the three-significant-figure approximation and by using the calculator in the following way.

$$\sin \theta° = \frac{23}{37}$$

$$\theta = 38.43.$$

In this version, the answer to $\frac{23}{37}$ was used directly to calculate the angle, and therefore all the figures were preserved in the process.

Sometimes it may be necessary to use a calculator memory to store an intermediate answer to as many figures as you need.

It is not necessary in this case, but you could calculate $\frac{23}{37}$ and put the result into memory A. Then you can calculate $\sin^{-1} A$ to get an accurate answer.

You need to be aware of this point for the multistage problems in Section 3.8, and especially so in Chapter 07.

3.6 Slope and gradient

Figure 3.10 represents a side view of the section of a rising path AC. AB is horizontal and BC is the vertical rise.

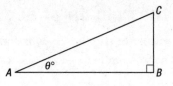

figure 3.10

Let the angle between the path and the horizontal be $\theta°$.

Then $\theta°$ is called the **angle of slope** or simply the **slope** of the path.

The ratio $\tan\theta°$ is called the **gradient** of the path.

Sometimes, especially by the side of railways, the gradient is given in the form 1 in 55. This means

$$\text{gradient} = \frac{1}{55}.$$

When the angle of slope $\theta°$ is very small, as in the case of a railway and most roads, it makes little practical difference if you take $\sin\theta°$ instead of $\tan\theta°$ as the gradient.

In practice also it is easier to measure $\sin\theta°$ (by measuring BC and AC), and the difference between AC and AB is relatively small provided that $\theta°$ is small.

You can use your calculator to see just how small the difference is between sines and tangents for small angles.

3.7 Projections

In Figure 3.11, let l be a straight line, and let AB be a straight line segment which makes an angle $\theta°$ with l.

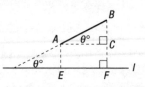

figure 3.11

Perpendiculars are drawn from A and B to l, meeting l at E and F. Then EF is called the **projection** of AB on l.

You can see from Figure 3.11 that the lengths AC and EF are equal.

As $\qquad\cos\theta° = \dfrac{AC}{AB},$

$$AC = AB\cos\theta°$$

so $\qquad\qquad EF = AB\cos\theta°.$

Exercise 3.2

This is a miscellaneous exercise involving sines, cosines and tangents.

In questions 1 to 10, find the marked angle or side.

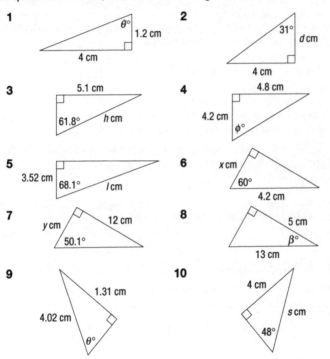

1 $\theta°$ 1.2 cm, 4 cm

2 31° d cm, 4 cm

3 5.1 cm, 61.8° h cm

4 4.8 cm, 4.2 cm $\phi°$

5 3.52 cm, 68.1° l cm

6 x cm, 60°, 4.2 cm

7 y cm, 12 cm, 50.1°

8 5 cm, $\beta°$, 13 cm

9 1.31 cm, 4.02 cm, $\theta°$

10 4 cm, s cm, 48°

11 In a right-angled triangle the two sides containing the right angle have lengths 2.34 m and 1.64 m. Find the smallest angle and the length of the hypotenuse.

12 In the triangle ABC, C is a right angle, AC is 122 cm and AB is 175 cm. Calculate angle B.

13 In triangle ABC, angle $C = 90°$, and $A = 37.35°$ and $AB = 91.4$ mm. Find the lengths of BC and CA.

14 ABC is a triangle. Angle C is a right angle, AC is 21.32 m and BC is 12.56 m. Find the angles A and B.

15 In a triangle *ABC*, *AD* is the perpendicular from *A* to *BC*. The lengths of *AB* and *BC* are 3.25 cm and 4.68 cm and angle *B* is 55°. Find the lengths of *AD*, *BD* and *AC*.

16 *ABC* is a triangle, right-angled at *C*. The lengths of *BC* and *AB* are 378 mm and 543 mm. Find angle *A* and the length of *CA*.

17 A ladder 20 m long rests against a vertical wall. Find the inclination of the ladder to the horizontal when the foot of the ladder is 7 m from the wall.

18 A ship starting from *O* travels 18 km h^{-1} in the direction 35° north of east. How far will it be north and east of *O* after an hour?

19 A pendulum of length 20 cm swings on either side of the vertical through an angle of 15°. Through what height does the bob rise?

20 The side of an equilateral triangle is *x* metres. Find in terms of *x* the altitude of the triangle. Hence find sin 60° and sin 30°.

21 A straight line 3.5 cm long makes an angle of 42° with the *x*-axis. Find the lengths of its projections on the *x*- and *y*-axes.

22 When you walk 1.5 km up the line of greatest slope of a hill you rise 94 m. Find the gradient of the hill.

23 A ship starts from a given point and sails 15.5 km on a bearing of 319°. How far has it gone west and north respectively?

24 A point *P* is 14.5 km north of *Q* and *Q* is 9 km west of *R*. Find the bearing of *P* from *R* and its distance from *R*.

3.8 Multistage problems

This section gives examples of multistage problems where you need to think out a strategy before you start.

You need also to be aware of the advice given in Section 3.5 about the accurate use of a calculator.

Example 3.5
ABCD is a kite in which *AB* = *AD* = 5 cm and *BC* = *CD* = 7 cm. Angle *DAB* = 80°. Calculate angle *BCD*.

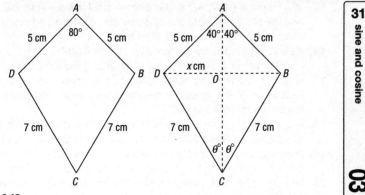

figure 3.12

The left-hand diagram in Figure 3.12 shows the information. In the right-hand diagram, the diagonals, which cut at right-angles at O, have been drawn, and the line AC, which is an axis of symmetry, bisects the kite.

Let DO be x cm, and let angle $DCA = \theta°$.

Then $$\sin 40° = \frac{x}{5}$$

so $$x = 5 \sin 40°.$$

Also $$\sin \theta° = \frac{x}{7}.$$

Substituting for x $$\sin \theta° = \frac{5 \sin 40°}{7}$$

so $$\theta = 27.33\ldots .$$

Angle $BCD = 2\theta° = 54.66°$, correct to two decimal places. ∎

Example 3.6

Figure 3.13 represents part of a symmetrical roof frame. $PA = 28$ m, $AB = 6$ m and angle $OPA = 21°$. Find the lengths of OP and OA.

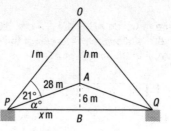

figure 3.13

Let $OP = l$ m, $PB = x$ m and $OA = h$ m.

To find l you need to find angle OPB; to do this you need first to find angle APB. Let angle $APB = \alpha°$.

Then $$\sin \alpha° = \frac{6}{30} = 0.21428...$$

so $$\alpha = 12.373.... .$$

Therefore angle $OPB = \alpha° + 21°$

$$= 33.373...°.$$

Next you must find the length x.

To find x, use Pythagoras's theorem in triangle APB.

$$x^2 = 28^2 - 6^2 = 784 - 36 = 748$$

so $$x = 27.349.... .$$

Then $$\cos 33.373...° = \frac{27.349...}{l}$$

so $$l = \frac{27.349...}{\cos 33.373...°}$$

$$= 32.7500.... .$$

To find h, you need to use $h = OB - AB = OB - 6$.

To find OB, $\sin 33.373...° = \dfrac{OB}{32.7500...}$

so $$OB = 32.7500... \times \sin 33.373...°$$

$$= 18.0156... ,$$

and $$h = 18.0156... - 6 = 12.0156.... .$$

Therefore $OP = 32.75$ m and $OA = 12.02$ m approximately. ∎

Exercise 3.3

1 *ABCD* is a kite in which *AB* = *AD* = 5 cm, *BC* = *CD* = 7 cm and angle *DAB* = 80°. Calculate the length of the diagonal *AC*.

2 In the diagram, find the length of *AC*.

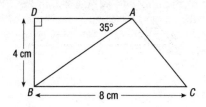

3 In the diagram, find the angle *θ*.

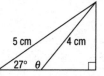

4 *PQRS* is a rectangle. A semicircle drawn with *PQ* as diameter cuts *RS* at *A* and *B*. The length *PQ* is 10 cm, and angle *BQP* is 30°. Calculate the length *PS*.

5 A ship sails 5 km on a bearing of 045° and then 6 km on a bearing of 060°. Find its distance and bearing from its starting point.

6 The lengths *AB* and *AC* of a triangle *ABC* are 5 cm and 6 cm respectively. The length of the perpendicular from *A* to *BC* is 4 cm. Calculate the angle *BAC*.

7 In a triangle *ABC*, the angles at *A* and *C* are 20° and 30° respectively. The length of the perpendicular from *B* to *AC* is 10 cm. Calculate the length of *AC*.

8 In the isosceles triangle *ABC*, the equal angles at *B* and *C* are each 50°. The sides of the triangle each touch a circle of radius 2 cm.

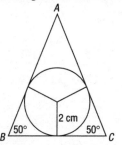

Calculate the length *BC*.

9 A ladder of length 5 m is leaning against a vertical wall at an angle of 60° to the horizontal. The foot of the ladder moves in by 50 cm. By how much does the top of the ladder move up the wall?

10 *PQRS* is a trapezium, with *PQ* parallel to *RS*. The angles at *P* and *Q* are 120° and 130° respectively. The length *PQ* is 6.23 cm and the distance between the parallel sides is 4.92 cm. Calculate the length of *RS*.

04

in three dimensions

In this chapter you will learn:
- the importance of good diagrams in solving three-dimensional problems
- how to break down three-dimensional problems into two-dimensional problems
- how to solve problems by using pyramids, boxes and wedges.

4.1 Introduction

Working in three dimensions introduces no new trigonometric ideas, but you do need to be able to think in three dimensions and to be able to visualize the problem clearly. To do this it is a great help to be able to draw a good figure.

You can solve all the problems in this chapter by picking out right-angled triangles from a three-dimensional figure, drawn, of course, in two dimensions.

This chapter will consist mainly of examples, which will include certain types of diagram which you should be able to draw quickly and easily.

4.2 Pyramid problems

The first type of diagram is the pyramid diagram. This diagram will work for all problems which involve pyramids with a square base.

Example 4.1

$ABCD$ is the square base of side 5 cm of a pyramid whose vertex V is 6 cm directly above the centre O of the square. Calculate the angle AVC.

The diagram is drawn in Figure 4.1.

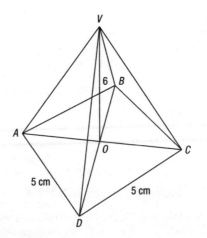

figure 4.1

There are a number of features you should notice about this diagram.

- The diagram is large enough to avoid points being 'on top of one another'.

- The vertex *V* should be above the centre of the base. It is best to draw the base first, draw the diagonals intersecting at *O*, and then put *V* vertically above *O*, siting *V* so that the edge *VC* of the pyramid does not appear to pass through *B*.

- Note that the dimensions have not been put on the measurements where, to do so, would add clutter.

To solve the problem, you must develop a strategy which involves creating or recognizing right-angled triangles.

The angle *AVC* is the vertical angle of the isosceles triangle *AVC* which is bisected by *VO*. If you can find the length *AO*, you can find angle *AVC*. But you can find *AO* by using the fact that *ABCD* is a square. See Figure 4.2.

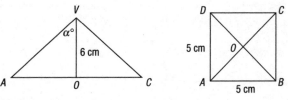

figure 4.2

Find *AC* by using Pythagoras's theorem

$$AC^2 = 5^2 + 5^2 = 50,$$

so $AC = \sqrt{50}$ cm and $AO = \frac{1}{2}\sqrt{50}$ cm.

Let angle *AVO* = $\alpha°$.

Then $$\tan\alpha° = \frac{\frac{1}{2}\sqrt{50}}{6}$$

and $$\alpha = 30.5089\ldots .$$

Therefore angle *AVC* = $2\alpha° = 61.02°$. ∎

If you find that you can solve the problem without drawing the subsidiary diagram in Figure 4.2, then do so. But many people find that it helps to see what is happening.

All square-based pyramid problems can be solved using this diagram.

Example 4.2

PQRSV is a pyramid with vertex *V* which is situated symmetrically above the mid-point *O* of the rectangular base *PQRS*. The lengths of *PS* and *RS* are 6 cm and 5 cm, and the slant height *VP* is 8 cm. Find the angle that the edge *VP* makes with the ground.

Figure 4.3 shows the situation.

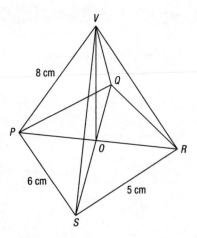

figure 4.3

Let the required angle *VPO* be $\alpha°$. Then you can find α from triangle *VPO* if you can find either *PO* or *VO*. You can find *PO* from the rectangle *PQRS*.

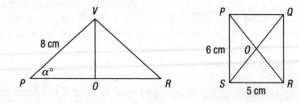

figure 4.4

As $PO = \frac{1}{2}PR$, and, using Pythagoras's theorem gives

$$PR = \sqrt{PS^2 + RS^2} = \sqrt{6^2 + 5^2} = \sqrt{61},$$

you find

$$PO = \tfrac{1}{2}\sqrt{61}\,\text{cm}.$$

Then

$$\cos\alpha° = \frac{PO}{VP} = \frac{\frac{1}{2}\sqrt{61}}{8},$$

so

$$\alpha = 60.78.$$

Therefore the edge VP makes an angle of 60.78° with the horizontal. ∎

4.3 Box problems

The second type of problem involves drawing a box.

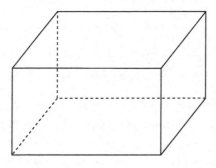

figure 4.5

The usual way to draw the box is to draw two identical rectangles, one 'behind' the other, and then to join up the corners appropriately. It can be useful to make some of the lines dotted to make it clear which face is in front.

Example 4.3

A room has length 4 m, width 3 m and height 2 m. Find the angle that a diagonal of the room makes with the floor.

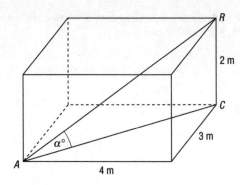

figure 4.6

Let the diagonal of the room be AR, and let AC be the diagonal of the floor. Let the required angle be $\alpha°$.

You can use Pythagoras's theorem to find the diagonal AC of the floor and so find the angle $\alpha°$.

$$AC = \sqrt{3^2 + 4^2} = \sqrt{25} = 5$$

so

$$\tan\alpha° = \frac{2}{5}$$

and

$$\alpha = 21.80.$$

Therefore the diagonal makes an angle of $21.80°$ with the floor. ∎

Sometimes the problem is a box problem, but may not sound like one.

Example 4.4

The top, P, of a pylon standing on level ground subtends an angle of $10°$ at a point S which is 50 m due south, and $5°$ at a point W lying west of the pylon. Calculate the distance SW correct to the nearest metre.

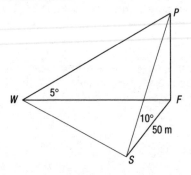

figure 4.7

This figure is part of the box diagram in Figure 4.5, with some lines removed. F is the foot of the pylon.

To calculate SW, start by finding the height of the pylon, then the length FW, and then use Pythagoras's theorem to find SW.

In triangle PFS, $\quad \tan 10° = \dfrac{PF}{50}$ so $PF = 50 \tan 10°$.

In triangle PFW $\quad \tan 5° = \dfrac{PF}{FW} = \dfrac{50 \tan 10°}{FW}$

so $\qquad\qquad FW = \dfrac{50 \tan 10°}{\tan 5°}.$

Finally $\qquad SW^2 = FW^2 + SF^2 = \left(\dfrac{50 \tan 10°}{\tan 5°}\right)^2 + 50^2$

$\qquad\qquad\qquad = 12654.86,$

so $\qquad SW = 112\,\text{m}$, correct to the nearest metre. ∎

4.4 Wedge problems

The third type of problem involves drawing a wedge. This wedge is really only part of a box, so you could think of a wedge problem as a special case of a box problem, but it is easier to think of it in a separate category.

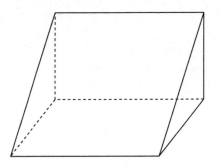

figure 4.8

Example 4.5

The line of greatest slope of a flat hillside slopes at an angle of 20° to the horizontal. To reduce the angle of climb, a walker walks on a path on the hillside which makes an angle of 50° with the line of greatest slope. At what angle to the horizontal does the walker climb on this path?

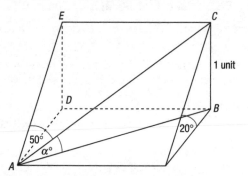

figure 4.9

Let AE be the line of greatest slope, and AC be the path of the walker. The angle that you need to find is angle BAC. Call this angle $\alpha°$. As there are no units to the problem, let the height BC be 1 unit.

To find α you need to find another length (apart from BC) in triangle BAC. This will come first from the right-angled triangle DAE, and then from triangle AEC, shown in Figure 4.10. Note that angle $DAE = 20°$ as AE is a line of greatest slope.

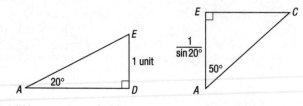

figure 4.10

In triangle AED $\qquad \sin 20° = \dfrac{1}{AE}$ so $AE = \dfrac{1}{\sin 20°}$.

In triangle EAC $\qquad \cos 50° = \dfrac{EA}{AC} = \dfrac{\frac{1}{\sin 20°}}{AC},$

so
$$AC = \frac{1}{\sin 20° \cos 50°}.$$

In triangle ABC $\sin \alpha = \frac{1}{AC} = \frac{1}{\dfrac{1}{\sin 20° \cos 50°}} = \sin 20° \cos 50°$

so
$$\alpha = 12.70. \blacksquare$$

The walker walks at 12.7° approximately to the line of greatest slope.

You may find it easier to follow if you evaluate AE and AC as you go along, but it is better practice to avoid it if you can.

Exercise 4

1 A pyramid has its vertex directly above the centre of its square base. The edges of the base are each 8 cm, and the vertical height is 10 cm. Find the angle between the slant face and the base, and the angle between a slant edge and the base.

2 A symmetrical pyramid stands on a square base of side 8 cm. The slant height of the pyramid is 20 cm. Find the angle between the slant edge and the base, and the angle between a slant face and the base.

3 A square board is suspended horizontally by four equal ropes attached to a point P directly above the centre of the board. Each rope has length 15 m and is inclined at an angle of 10° with the vertical. Calculate the length of the side of the square board.

4 A pyramid has its vertex directly above the centre of its square base. The edges of the base are each 6 cm, and the vertical height is 8 cm. Find the angle between two adjacent slant faces.

5 Find the angle that a main diagonal of a cube makes with the base. (Assume that the cube has sides of length 1 unit.)

6 A pylon is situated at a corner of a rectangular field with dimensions 100 m by 80 m. The angle subtended by the pylon at the furthest corner of the field is 10°. Find the angles subtended by the pylon at the other two corners of the field.

7 A regular tetrahedron has all its edges 8 cm in length. Find the angles which an edge makes with the base.

8 All the faces of a square-based pyramid of side 6 cm slope at an of 60° to the horizontal. Find the height of the pyramid, and the angle between a sloping edge and the base.

9 A vertical flag pole standing on horizontal ground has six ropes attached to it at a point 6 m from the ground. The other ends of the ropes are attached to points on the ground which lie in a regular hexagon with sides 4 m. Find the angle which a rope makes with the ground.

10 The diagram shows a roof structure. *PQRS* is a horizontal rectangle. The faces *ABRQ*, *ABSP*, *APQ* and *BRS* all make an angle of 45° with the horizontal.

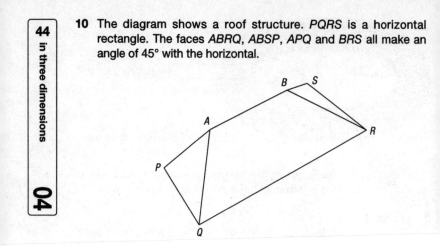

Find the angle made by the sloping edges with the horizontal.

angles of any magnitude

In this chapter you will learn:
- how to extend the definitions of sine, cosine and tangent to angles greater than 90° and less than 0°
- the shape of the graphs of sine, cosine and tangent
- the meaning of the terms 'period' and 'periodic'.

5.1 Introduction

If, either by accident or by experimenting with your calculator, you have tried to find sines, cosines and tangents of angles outside the range from 0 to 90°, you will have found that your calculator gives you a value. But what does this value mean, and how is it used? This chapter gives some answers to those questions.

If you think of sine, cosine and tangent only in terms of ratios of 'opposite', 'adjacent' and 'hypotenuse', then it is difficult to give meanings to trigonometric ratios of angles outside the interval 0 to 180° – after all, angles of triangles have to lie within this interval. However, the definition of tangent given in Section 2.3 and the definition of sine and cosine in Section 3.2 both extend naturally to angles of any magnitude.

It is convenient to start with the sine and cosine.

5.2 Sine and cosine for any angle

In Section 3.2, the following construction was given as the basis of the definition of sine and cosine.

Draw a circle with radius 1 unit, and centre at the origin O. Draw the radius OP at an angle $\theta°$ to the x-axis in an anti-clockwise direction. See Figure 5.1. Let P have coordinates (x, y).

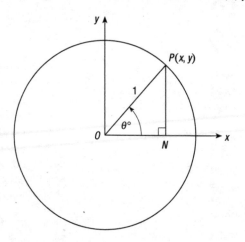

figure 5.1

Then $\sin\theta° = y$ and $\cos\theta° = x$ are the definitions of sine and cosine for any size of the angle $\theta°$.

The arrow labelling the angle $\theta°$ in Figure 5.1 emphasizes that angles are measured positively in the anti-clockwise direction, and negatively in the clockwise direction.

It is useful to divide the plane into four quadrants, called 1, 2, 3 and 4, as shown in Figure 5.2.

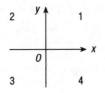

figure 5.2

Then for any given angles such as 60°, 210° and –140°, you can associate a quadrant, namely, the quadrant in which the radius corresponding to the angle lies.

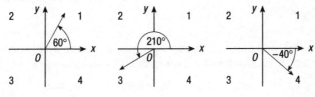

figure 5.3

In Figure 5.3, you can see that 60° is in quadrant 1, and is called a first quadrant angle; 210° is a third quadrant angle; –40° is a fourth quadrant angle.

You can have angles greater than 360°. For example, you can check that 460° is a second quadrant angle, and –460° is a third quadrant angle.

The definition of $\sin\theta°$, namely $\sin\theta° = y$, shows that if $\theta°$ is a first or second quadrant angle, then the y-coordinate of P is positive so $\sin\theta° > 0$; if $\theta°$ is a third or fourth quadrant angle, $\sin\theta° < 0$.

You can see from this definition, and from Figure 5.3, that $\sin 60° > 0$, that $\sin 210° < 0$ and that $\sin(-40°) < 0$; you can easily check these from your calculator.

Similarly, the definition of $\cos\theta°$, namely $\cos\theta° = x$, shows that if $\theta°$ is a first or fourth quadrant angle then the x-coordinate of P is positive so $\cos\theta° > 0$; if $\theta°$ is a second or third quadrant angle, $\cos\theta° < 0$.

You can also see that $\cos 60° > 0$, that $\cos 210° < 0$ and that $\cos(-40°) > 0$. Again, you can easily check these from your calculator.

Sine and cosine for multiples of 90°

The easiest way to find the sine and cosine of angles such as 90°, 540° and –90° is to return to the definitions, that is $\sin\theta° = y$ and $\cos\theta° = x$. See Figure 5.4.

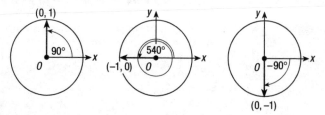

figure 5.4

Then you see from the left-hand diagram that the radius for 90° ends at (0, 1), so $\sin 90° = 1$ and $\cos 90° = 0$.

Similarly, the radius for 540° ends up at (–1, 0), so $\sin 540° = 0$ and $\cos 540° = -1$.

Finally, the radius for –90° ends up at (–1, 0), so $\sin(-90°) = -1$ and $\cos(-90°) = 0$.

Once again, you can check these results from your calculator.

5.3 Graphs of sine and cosine functions

As the sine and cosine functions are defined for all angles you can draw their graphs.

Figure 5.5 shows the graph of $y = \sin\theta°$ drawn for values of θ from –90 to 360.

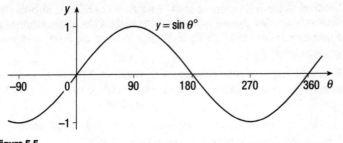

figure 5.5

You can see that the graph of $y = \sin \theta°$ has the form of a wave. As it repeats itself every 360°, it is said to be **periodic**, with **period** 360°. As you would expect from Section 5.2, the value of $\sin \theta°$ is positive for first and second quadrant angles and negative for third and fourth quadrant angles.

Figure 5.6 shows the graph of $y = \cos \theta°$ drawn for values of θ from −90 to 360.

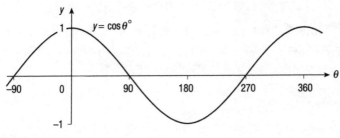

figure 5.6

As you can see, the graph of $y = \cos \theta°$ also has the form of a wave. It is also **periodic**, with **period** 360°. The value of $\cos \theta°$ is positive for first and fourth quadrant angles and negative for second and third quadrant angles.

It is the wave form of these graphs and their periodic properties which make the sine and cosine so useful in applications. This point is taken further in physics and engineering.

Exercise 5.1

In questions 1 to 8, use your calculator to find the following sines and cosines.

1 sin 130°	**2** cos 140°
3 sin 250°	**4** cos 370°
5 sin(−20)°	**6** cos 1000°
7 sin 36000°	**8** cos(−90)°

In questions 9 to 14, say in which quadrant the given angle lies.

9 200°	**10** 370°
11 (−300)°	**12** 730°
13 −600°	**14** 1000°

In questions 15 to 20, find the following sines and cosines without using your calculator.

15 cos 0	**16** sin 180°
17 cos 270°	**18** sin(−90)°
19 cos(−180)°	**20** sin 450°

5.4 The tangent of any angle

In Section 2.3 you saw that the definition of the tangent for an acute angle was given by $\tan \theta° = \dfrac{y}{x}$. This definition is extended to all angles, positive and negative. See Figure 5.7.

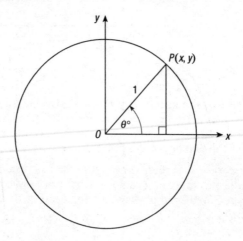

figure 5.7

If the angle $\theta°$ is a first quadrant angle, $\tan\theta°$ is positive. For a second quadrant angle, y is positive and x is negative, so $\tan\theta°$ is negative. For a third quadrant angle, y and x are both negative, so $\tan\theta°$ is positive. And for a fourth quadrant angle, y is negative and x is positive, so $\tan\theta°$ is negative.

5.5 Graph of the tangent function

Just as you can draw graphs of the sine and cosine functions, you can draw a graph of the tangent function. Its graph is shown in Figure 5.8.

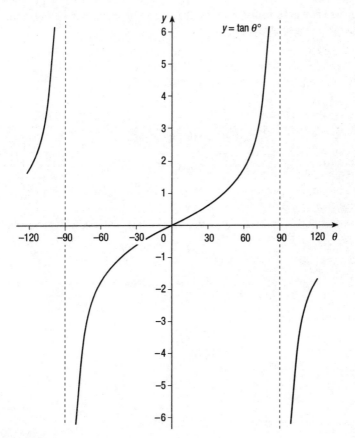

figure 5.8

You can see from Figure 5.8 that, like the sine and cosine functions, the tangent function is periodic, but with period 180°, rather than 360°.

You can also see that, for odd multiples of 90°, the tangent function is not defined. You cannot talk about tan 90°. It does not exist.

5.6 Sine, cosine and tangent

There is an important relation between the sine, cosine and tangent which you can deduce immediately from their definitions.

From the definitions

$$\sin \theta° = y,$$
$$\cos \theta° = x,$$
$$\tan \theta° = \frac{y}{x},$$

you can see that

$$\tan \theta° = \frac{\sin \theta°}{\cos \theta°}.$$ **Equation 1**

Equation 1 will be used repeatedly throughout the remainder of the book.

Exercise 5.2

In questions 1 to 4, use your calculator to find the following tangents.

1   tan 120°   2   tan(–30)°
3   tan 200°   4   tan 1000°

5   Attempt to find tan 90° on your calculator. You should find that it gives some kind of error message.

6   Calculate the value of $\frac{\sin 12°}{\cos 12°}$.

7   Calculate the value of $\frac{\cos 1000°}{\sin 1000°}$.

06

solving simple equations

In this chapter you will learn:
- how to solve simple equations involving sine, cosine and tangent
- the meaning of 'principal angle'
- how to use the principal angle to find all solutions of the equation.

6.1 Introduction

This chapter is about solving equations of the type $\sin\theta° = 0.4$, $\cos\theta° = 0.2$ and $\tan\theta° = 0.3$.

It is easy, using a calculator, to find the sine of a given angle. It is also easy, with a calculator, to find one solution of an equation such as $\sin\theta° = 0.4$. You use the $\sin^{-1}$ key and find $\theta = 23.57\ldots$. So far so good.

The problem is that Figure 6.1 shows there are many angles, infinitely many in fact, for which $\sin\theta° = 0.4$. You have found one of them – how do you find the others from the angle that you have found?

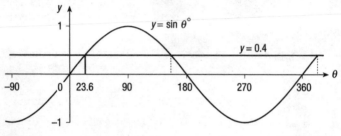

figure 6.1

Figure 6.1 shows that there is another angle lying between 90° and 180° satisfying the equation $\sin\theta° = 0.4$, and then infinitely many others, repeating every 360°.

6.2 Solving equations involving sines

Principal angles

The angle given by your calculator when you press the $\sin^{-1}$ key is called the **principal angles**.

For the sine function the principal angle lies in the interval $-90 < \theta \leqslant 90$.

If you draw the graph of $y = \sin^{-1} x$ using your calculator values you get the graph shown in Figure 6.2.

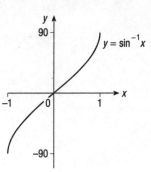

figure 6.2

So the question posed in Section 6.1 is, 'Given the principal angle for which $\sin\theta° = 0.4$, how do you find all the other angles?'

Look at the sine graph in Figure 6.1. Notice that it is symmetrical about the 90° point on the θ-axis. This shows that for any angle $\alpha°$

$$\sin(90 - \alpha)° = \sin(90 + \alpha)°.\qquad\textbf{Equation 1}$$

If you write $x = 90 - \alpha$, then $\alpha = 90 - x$, so $90 + \alpha = 180 - x$. Equation 1 then becomes, for any angle $\alpha°$

$$\sin\alpha° = \sin(180 - \alpha)°.\qquad\textbf{Equation 2}$$

Equation 2 is the key to solving equations which involve sines.

Returning to the graph of $y = \sin\theta°$ in Figure 6.1, and using Equation 2, you can see that the other angle between 0 and 180 with $\sin\theta° = 0.4$ is

$$180 - \sin^{-1}0.4 = 180 - 23.57... = 156.42... \,.$$

Now you can add (or subtract) multiples of 360° to find all the other angles solving $\sin\theta° = 0.4$, and obtain

$$\theta = 23.57,\ 156.42,\ 383.57,\ 516.42,...$$

correct to two decimal places.

Summary

To solve an equation of the form $\sin x° = c$ where c is given:

- find the principal angle
- use Equation 2 to find another angle for which $\sin x° = c$
- add or subtract any multiple of 360.

Example 6.1

Solve the equation $\sin x° = -0.2$, giving all solutions in the interval from -180 to 180.

The principal angle is -11.54.

From Equation 2, $180 - (-11.54) = 191.54$ is also a solution, but this is outside the required range. However, as you can add and subtract any multiple of 360, you can find the solution between -180 and 180 by subtracting 360.

Therefore the required solution is

$$191.54 - 360 = -168.46.$$

Therefore the solutions are -168.46 and -11.54. ∎

Example 6.2

Solve the equation $\sin 2x° = 0.5$, giving all solutions from 0 to 360.

Start by letting $y = 2x$. Then you have first to solve for y the equation $\sin y° = 0.5$. Note also that if x lies between 0 and 360, then y, which is $2x$, lies between 0 and 720.

The principal angle for the solution of $\sin y° = 0.5$ is $\sin^{-1} 0.5 = 30$.

From Equation 2, $180 - 30 = 150$ is also a solution.

Adding multiples of 360 shows that 390 and 510 are also solutions for y.

Thus $y = 2x = 30, 150, 390, 510$

so $x = 15, 75, 195, 255.$ ∎

Example 6.3

Find the smallest positive root of the equation $\sin(2x + 50)° = 0.1$.

Substitute $y = 2x + 50$, so you solve $\sin y° = 0.1$.

From $y = 2x + 50$ you find that $x = \frac{1}{2}(y - 50)$.

As $x > 0$ for a positive solution, $\frac{1}{2}(y - 50) > 0$, so $y > 50$.

The principal angle solving $\sin y° = 0.1$ is 5.74.

From Equation 2, $180 - 5.74 = 174.26$ is the first solution greater than 50.

As $x = \frac{1}{2}(y - 50)$,

$$x = \frac{1}{2}(174.26 - 50) = 62.13.$$ ∎

Exercise 6.1

In questions 1 to 8, find the solutions of the given equation in the interval from 0 to 360.

1 $\sin\theta° = 0.3$ **2** $\sin\theta° = 0.45$
3 $\sin\theta° = 0$ **4** $\sin\theta° = 1$
5 $\sin\theta° = -1$ **6** $\sin\theta° = -0.1$
7 $\sin\theta° = -0.45$ **8** $\sin\theta° = -0.5$

In questions 9 to 16, find the solutions of the given equation in the interval from −180 to 180.

9 $\sin\theta° = -0.15$ **10** $\sin\theta° = -0.5$
11 $\sin\theta° = 0$ **12** $\sin\theta° = 1$
13 $\sin\theta° = -1$ **14** $\sin\theta° = 0.9$
15 $\sin\theta° = -0.9$ **16** $\sin\theta° = -0.766$

In questions 17 to 24, find the solutions of the given equation in the interval from 0 to 360.

17 $\sin 2x° = 0.5$ **18** $\sin 2\theta° = 0.45$
19 $\sin 3\theta° = 0$ **20** $\sin 2\theta° = -1$
21 $\sin\frac{1}{2}\theta° = 0.5$ **22** $\sin\frac{1}{2}\theta° = 1$
23 $\sin 3\theta° = -0.5$ **24** $3\sin 2x° = 2$

25 The height h in metres of the water in a harbour t hours after the water is at its mean level is given by $h = 6 + 4\sin(30t)°$. Find the first positive value of t for which the height of the water first reaches 9 metres.

26 The length l hours of a day in hours t days after the beginning of the year is given approximately by

$$h = 12 - 6\cos\left(\frac{360}{365}t\right)°.$$

Find the approximate number of days per year that the length of day is longer than 15 hours.

6.3 Solving equations involving cosines

To solve an equation of the form $\cos\theta° = 0.2$ it is helpful to look at the graph of the cosine function in Figure 6.3.

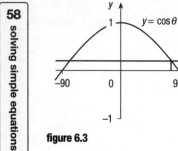
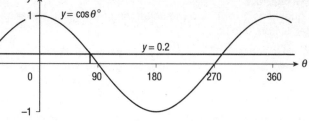

figure 6.3

For the cosine the principal angle is in the interval $0 \leqslant \theta \leqslant 180$. For the equation $\cos \theta° = 0.2$, the principal angle is $78.46\ldots$.

If you draw the graph of $y = \cos^{-1} x$ using your calculator

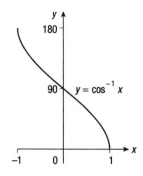

figure 6.4

values, you get the graph shown in Figure 6.4.

The symmetry of the cosine graph in Figure 6.3 shows that

$$\cos(-\theta°) = \cos \theta°. \qquad \textbf{Equation 3}$$

If you use Equation 3, you find that $\cos(-78.46\ldots)° = 0.2$.

If the interval for which you need the solution is from -180 to 180, you have two solutions, $-78.46\ldots$ and $78.46\ldots$.

If you need solutions between 0 and 360, you can add 360 to the first solution and obtain 78.46 and 281.54, correct to two decimal places.

Summary

To solve an equation of the form $\cos\theta° = c$ where c is given:

- find the principal angle
- use Equation 3 to find another angle for which $\cos\theta° = c$
- add or subtract any multiple of 360.

Example 6.4

Solve the equation $\cos\theta° = -0.1$ giving all solutions in the interval −180 to 180, correct to two decimal places.

For $\cos\theta° = -0.1$, the principal angle is 95.74... .

Using Equation 3, the other angle in the required interval is −95.74... .

The solutions, correct to two decimal places, are −95.74 and 95.74. ∎

Example 6.5

Find all the solutions of the equation $2\cos 3x° = 1$ in the interval 0 to 360.

The equation $2\cos 3x° = 1$ can be written in the form $\cos y° = \frac{1}{2}$, where $y = 3x$. If x lies in the interval 0 to 360, then y, which is $3x$, lies in the interval 0 to 1080.

The principal angle for the solution of $\cos y° = \frac{1}{2}$ is $\cos^{-1}\frac{1}{2} = 60$.

From Equation 3, −60 is also a solution of $\cos y° = \frac{1}{2}$.

Adding multiples of 360 shows that 300, 420, 660, 780 and 1020 are also solutions for y.

Thus $y = 3x = 60, 300, 420, 660, 780, 1020$

so $x = 20, 100, 140, 220, 260, 340.$ ∎

6.4 Solving equations involving tangents

To solve an equation of the form $\tan\theta° = 0.3$ it is helpful to look at the graph of the tangent function, Figure 5.8.

The symmetry of the tangent graph shows that

$$\tan\theta° = \tan(180 + \theta)°. \qquad \textbf{Equation 4}$$

For the tangent the principal angle is in the interval $-90 < \theta < 90$. For the equation $\tan\theta° = 0.3$, the principal angle is $16.70...$.

Using Equation 4 you find that the other angle between 0 and 360 satisfying $\tan\theta° = 0.3$ is $180 + 16.70... = 196.70...$.

Summary

To solve an equation of the form $\tan\theta° = c$ where c is given:

- find the principal angle
- use Equation 4 to find another angle for which $\tan\theta° = c$
- add or subtract any multiple of 360.

Example 6.6
Solve the equation $\tan\theta° = -0.6$ giving all solutions in the interval -180 to 180 correct to two decimal places.

For $\tan\theta° = -0.6$, the principal angle is $-30.96...$.

Using Equation 4, the other angle in the required interval, -180 to 180 is $180 + (-30.96...) = 149.04...$.

Therefore the solutions, correct to two decimal places, are -30.96 and 149.04. ∎

Example 6.7
Find all the solutions of the equation $\tan\frac{3}{2}\theta° = -1$ in the interval 0 to 360.

The equation $\tan\frac{3}{2}\theta° = -1$ can be written in the form $\tan y° = -1$, where $y = \frac{3}{2}\theta$. If θ lies in the interval 0 to 360, then y, which is $\frac{3}{2}\theta$, lies in the interval 0 to 540.

The principal angle for the solution of $\tan y° = -1$ is $\tan^{-1}(-1) = -45$.

From Equation 4, 135 is also a solution of $\tan y° = -1$.

Adding multiples of 360 shows that 135, 315 and 495 are also solutions for y in the interval from 0 to 540.

Thus $$y = \tfrac{3}{2}\theta = 135, 315, 495$$

so $$\theta = 90, 210, 330. \blacksquare$$

Exercise 6.2

In questions 1 to 10, find all the solutions to the given equation in the interval 0 to 360 inclusive.

1 $\cos\theta° = -\frac{1}{3}$

2 $\tan x° = 2$

3 $\cos\alpha° = \frac{3}{4}$

4 $\tan\beta° = -0.5$

5 $\cos 2\theta° = \frac{1}{2}$

6 $\tan 2\theta° = 1$

7 $\cos\frac{1}{2}\theta° = -0.2$

8 $\tan\frac{1}{3}x° = 1.1$

9 $\cos 2\theta° = -0.766$

10 $\tan 2x° = -0.1$

In questions 11 to 16, find all the solutions to the given equation in the interval −180 to 180 inclusive.

11 $\cos 2x° = -0.3$

12 $\tan 2x° = -0.5$

13 $\sin 2\theta° = 0.4$

14 $\cos\frac{2}{3}x° = 0.5$

15 $\tan\frac{3}{2}x° = 1$

16 $\sin\frac{2}{3}x° = -0.5$

17 The height h in metres of water in a harbour above low tide is given by the equation $h = 14 - 10\cos(30t)°$ where t is measured in hours from midday. A ship can enter the harbour when the water is greater than 20 metres. Between what times can the boat first enter the harbour?

07

**the sine and
cosine formulae**

In this chapter you will learn:

- how to solve problems in triangles which are not right-angled
- how to use the sine and cosine formulae for a triangle, and how to find the area of a triangle
- some applications of the sine and cosine formulae for a triangle.

7.1 Notation

This chapter is about finding the lengths of sides and the magnitudes of angles of triangles which are not right-angled. It is useful to have some notation about sides and angles of triangles.

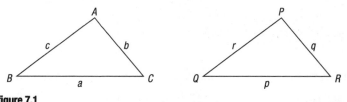

figure 7.1

In the triangle ABC, the angles are called A, B and C; the sides opposite these angles are given the corresponding lower-case letters, a, b and c. In the triangle PQR, the sides opposite the angles P, Q and R are p, q and r.

In this chapter, the angles will be measured in degrees: thus, in Figure 7.1, $A = 95°$ approximately.

The sides may be measured in any units you choose, provided they are all measured in the same units. In some of the diagrams that follow the units of length are omitted.

7.2 Area of a triangle

You are familiar with the formula

$$\tfrac{1}{2} \times \text{base} \times \text{height}$$

for the area of a triangle, whether it is acute-angled or obtuse-angled.

To find a formula for the area in terms of the sides and angles of a triangle, you need to consider acute- and obtuse-angled triangles separately.

Case 1: Acute-angled triangle

Figure 7.2 shows an acute-angled triangle with the perpendicular drawn from A onto BC. Let the length of this perpendicular be h.

Let the area of the triangle be Δ, so $\Delta = \tfrac{1}{2}ah$.

To find h you can use the left-hand right-angled triangle.

Thus, $$\sin B = \frac{h}{c},$$

so $$h = c\sin B.$$

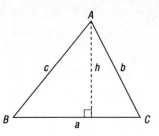

figure 7.2

Therefore $\Delta = \frac{1}{2}ah = \frac{1}{2}ac\sin B.$ **Equation 1**

Alternatively, you could find h from the other right-angled triangle,

$$\sin C = \frac{h}{b} \text{ so } h = b\sin C,$$

so $\Delta = \frac{1}{2}ah = \frac{1}{2}ab\sin C.$ **Equation 2**

So, using Equations 1 and 2 for the area of an acute-angled triangle,

$$\Delta = \frac{1}{2}ac\sin B = \frac{1}{2}ab\sin C. \qquad \textbf{Equation 3}$$

Notice the symmetry of the two expressions in Equation 3. Each of the expressions $\frac{1}{2}ac\sin B$ and $\frac{1}{2}ab\sin C$ consists of $\frac{1}{2}$ the product of two sides multiplied by the sine of the angle between them. It follows from this symmetry that there is a third expression for the area of the triangle, namely $\frac{1}{2}bc\sin A$.

You could have derived this expression directly if in Figure 7.2 you had drawn the perpendicular from B to AC, or from C to AB.

Thus for an acute-angled triangle,

$$\Delta = \frac{1}{2}bc\sin A = \frac{1}{2}ca\sin B = \frac{1}{2}ab\sin C.$$

Case 2: Obtuse-angled triangle

Figure 7.3 shows a triangle, obtuse-angled at C, with the perpendicular drawn from A onto BC. Let the length of this perpendicular be h.

Let the area of the triangle be Δ, so $\Delta = \frac{1}{2}ah$.

To find h you can use the larger right-angled triangle.

Thus, $\sin B = \frac{h}{c},$

so $h = c\sin B.$

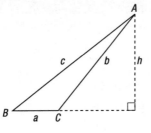

figure 7.3

Therefore $\Delta = \frac{1}{2}ah = \frac{1}{2}ac\sin B.$ **Equation 4**

Alternatively, you could find h from the other right-angled triangle,

$$\sin(180° - C) = \frac{h}{b} \text{ so } h = b\sin(180° - C),$$

so $\Delta = \frac{1}{2}ah = \frac{1}{2}ab\sin(180° - C).$ **Equation 5**

Equation 5 looks quite different from Equation 4, but if you recall Equation 2 from Chapter 06, you find

$$\sin C = \sin(180° - C).$$

Therefore you can write Equation 5 as

$$\Delta = \frac{1}{2}ab\sin(180° - C) = \frac{1}{2}ab\sin C.$$

Putting Equations 4 and 5 together,

$$\Delta = \frac{1}{2}ac\sin B = \frac{1}{2}ab\sin C.$$ **Equation 6**

Notice again the symmetry of the two expressions in Equation 6. Each of the expressions $\frac{1}{2}ac\sin B$ and $\frac{1}{2}ab\sin C$ consists of $\frac{1}{2}$ the product of two sides multiplied by the sine of the angle between them. It follows from this symmetry that there is a third expression for the area of the triangle, namely $\frac{1}{2}bc\sin A$.

Thus for an obtuse-angled triangle,

$$\Delta = \frac{1}{2}bc\sin A = \frac{1}{2}ca\sin B = \frac{1}{2}ab\sin C.$$

Case 3: Area of any triangle

Thus, the area Δ of any triangle, acute-angled or obtuse-angled, is

$$\Delta = \frac{1}{2}bc\sin A = \frac{1}{2}ca\sin B = \frac{1}{2}ab\sin C. \quad \textbf{Equation 7}$$

If you think of this formula as

$$\text{Area} = \frac{1}{2} \times \text{product of sides} \times \text{sine of included angle}$$

then it is independent of the lettering of the particular triangle involved. The units of area will be dependent on the unit of length that is used.

Example 7.1
Find the area of the triangle with $a = 12\,$cm, $b = 11\,$cm and $C = 53°$.

The area $\Delta\,$cm^2 of the triangle is

$$\Delta = \tfrac{1}{2}ab\sin C$$
$$= \tfrac{1}{2} \times 12 \times 11 \times \sin 53°$$
$$= 52.71.$$

The area of the triangle is $52.7\,$cm^2. ∎

Example 7.2
Find the area of the triangle with $b = 6\,$cm, $c = 4\,$cm and $A = 123°$.

The area $\Delta\,$cm^2 of the triangle is

$$\Delta = \tfrac{1}{2}bc\sin A$$
$$= \tfrac{1}{2} \times 6 \times 4 \times \sin 123°$$
$$= 10.06.$$

The area of the triangle is $10.1\,$cm^2. ∎

7.3 The sine formula for a triangle

The formula for the area of a triangle,

$$\Delta = \tfrac{1}{2}bc\sin A = \tfrac{1}{2}ca\sin B = \tfrac{1}{2}ab\sin C$$

leads to a very important result.

Leave out the Δ, and multiply the resulting equation by 2, and you find

$$bc\sin A = ca\sin B = ab\sin C.$$

If you now divide by the product abc, you obtain

$$\frac{\sin A}{a} = \frac{\sin B}{b} = \frac{\sin C}{c}. \qquad \textbf{Equation 8}$$

This formula is called **the sine formula for a triangle,** or more briefly, **the sine formula.**

Example 7.3
In triangle ABC in Figure 7.4, angle $A = 40°$, angle $B = 80°$ and $b = 5\,$cm. Find the length of the side a.

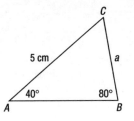

figure 7.4

Using the sine formula,

$$\frac{\sin 40°}{a} = \frac{\sin 80°}{5},$$

so
$$a = \frac{5 \sin 40°}{\sin 80°}$$

$$= 3.26.$$

The length of side a is 3.26 cm, correct to three significant figures. ∎

Sometimes you may need to attack a problem indirectly.

Example 7.4

In triangle ABC in Figure 7.5, angle $C = 100°$, $b = 4$ cm and $c = 5$ cm. Find the magnitude of the angle B.

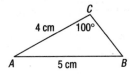

figure 7.5

Using the sine formula,

$$\frac{\sin B}{4} = \frac{\sin 100°}{5}$$

so
$$\sin B = \frac{4 \sin 100°}{5}.$$

This is an equation in $\sin B$; the principal angle is 51.98°.

The two angles between 0 and 180° satisfying this equation are 51.98° and 128.02°. However, as the angles of the triangle add up to 180°, and one of the angles is 100°, the other angles must be acute.

Therefore $\qquad B = 51.98°.$

Angle B is $51.98°$ correct to two decimal places. ∎

The situation that arose in Example 7.4 where you needed to think carefully about the two angles which satisfy an equation of the form $\sin\theta° = \ldots$ occurs in other cases. You cannot always resolve which solution you need.

Example 7.5 shows an example of this kind.

7.4 The ambiguous case

Example 7.5

In a triangle ABC, angle $B = 40°$, $a = 5$ cm and $b = 4$ cm. Find the angle A.

Using the sine formula,

$$\frac{\sin A}{5} = \frac{\sin 40°}{4}$$

so $\qquad \sin A = \dfrac{5\sin 40°}{4}.$

This is an equation in $\sin A$; the principal angle is $53.46°$.

The two angles between 0 and $180°$ satisfying this equation are $53.46°$ and $126.54°$. ∎

This time however, there is no reason to rule out the obtuse case, so there are two possible answers to the question.

Angle A is $53.46°$ or $126.54°$ correct to two decimal places.

Figure 7.6 shows what is happening.

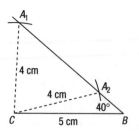

figure 7.6

If you try to draw the triangle with the information given, you would draw the side a of length 5 cm, and draw the 40° at B. If you then draw a circle of radius 4 cm with its centre at C you find that it cuts the line which you drew through B in two places, A_1 and A_2.

One triangle is the acute-angled A_1BC, where angle $A_1 = 53.46°$. The other triangle is the obtuse-angled triangle A_2BC where $A_2 = 126.54°$.

This situation in which the triangle is not defined uniquely is called the **ambiguous case**. It shows that when you use the sine formula to find an angle it is important to take account of both solutions of the equation $\sin \theta° = ...$ which arises in the solution.

Exercise 7.1

In questions 1 to 5, find the unknown sides of the triangle ABC, and calculate the area of the triangle.

1   $A = 54°$, $B = 67°$ and $a = 13.9$ cm.
2   $A = 38.25°$, $B = 29.63°$ and $b = 16.2$ cm
3   $A = 70°$, $C = 58.27°$ and $b = 6$ mm
4   $A = 88°$, $B = 36°$ and $a = 9.5$ cm
5   $B = 75°$, $C = 42°$ and $b = 25.0$ cm

In questions 6 to 9, there may be more than one solution. Find all the solutions possible.

6   $b = 30.4$ cm, $c = 34.8$ cm, $B = 25°$. Find C, A and a.
7   $b = 70.25$ cm, $c = 85.3$ cm, $B = 40°$. Find C, A and a.
8   $a = 96$ cm, $c = 100$ cm, $C = 66°$. Find A, B and b.
9   $a = 91$ cm, $c = 78$ cm, $C = 29.45°$. Find A, B and b.

7.5 The cosine formula for a triangle

The sine formula is easy to remember and easy to use, but it is no help if you know the lengths of two sides and the included angle and wish to find the length of the third side.

In Figure 7.7, suppose that you know the lengths of the sides a and b, and the angle C between them, and that you wish to find the length of the side c.

There are two cases to consider: when angle C is acute and when angle C is obtuse. In both cases, the perpendicular from A is drawn, meeting BC at D. The length of this perpendicular is h. Let the length CD be x.

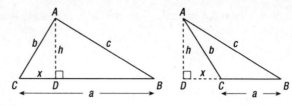

figure 7.7

Case 1: Acute-angled triangle

Using Pythagoras's theorem in the triangle ACD,

$$h^2 = b^2 - x^2.$$

Using Pythagoras's theorem in the triangle ABD,

$$h^2 = c^2 - (a-x)^2.$$

Equating these expressions for h^2 gives

$$b^2 - x^2 = c^2 - (a-x)^2,$$

that is

$$b^2 - x^2 = c^2 - a^2 + 2ax - x^2$$

or

$$c^2 = a^2 + b^2 - 2ax.$$

In triangle ACD, $x = b \cos C$, so the expression for c^2 becomes

$$c^2 = a^2 + b^2 - 2ab \cos C.$$

Case 2: Obtuse-angled triangle

Using Pythagoras's theorem in the triangle ACD,

$$h^2 = b^2 - x^2.$$

Using Pythagoras's theorem in the triangle ABD,

$$h^2 = c^2 - (a+x)^2.$$

Equating these expressions for h^2 gives

$$b^2 - x^2 = c^2 - (a+x)^2,$$

that is

$$b^2 - x^2 = c^2 - a^2 - 2ax - x^2$$

or

$$c^2 = a^2 + b^2 + 2ax.$$

In triangle ACD, $x = b \cos(180° - C)$. However, from the graph of $y = \cos\theta°$ in Section 6.3, $\cos(180° - C) = -\cos C$. Thus when C is obtuse, $x = -b \cos C$. The expression for c^2 therefore becomes

$$c^2 = a^2 + b^2 - 2ab \cos C.$$

This formula is called **the cosine formula for a triangle**, or more briefly, **the cosine formula**.

Notice that the formula $c^2 = a^2 + b^2 - 2ab \cos C$ is symmetrical in the letters, a, b, c, and A, B and C. There are two other formulae like it. These three formulae are

$$\left.\begin{aligned} a^2 &= b^2 + c^2 - 2bc \cos A \\ b^2 &= c^2 + a^2 - 2ca \cos B \\ c^2 &= a^2 + b^2 - 2ab \cos C \end{aligned}\right\} \qquad \text{Equations 9}$$

Example 7.6
In a triangle, a = 4 cm, b = 7 cm and angle C = 73°. Find the length of the side c.

Using the cosine formula,

$$c^2 = a^2 + b^2 - 2ab \cos C$$

gives $\qquad\qquad c^2 = 4^2 + 7^2 - 2 \times 4 \times 7 \cos 73°$

and $\qquad\qquad c = 6.97.$

The length of the side c is 6.97 cm approximately. ∎

You can also use the cosine formula to find an unknown angle if you know the lengths of the three sides.

Example 7.7
The three sides of a triangle have lengths 5 cm, 4 cm and 8 cm. Find the largest angle of the triangle.

The largest angle is the angle opposite the longest side. Using the cosine formula and calling the angle $\theta°$, you find

$$8^2 = 5^2 + 4^2 - 2 \times 5 \times 4 \cos \theta°$$

$$64 = 25 + 16 - 40 \cos \theta°$$

$$\cos \theta° = -\frac{23}{40}$$

$$\theta = 125.10.$$

The largest angle of the triangle is 125.10° approximately. ∎

Notice that there is no ambiguity in using the cosine formula. If the angle is obtuse, the cosine will be negative.

Exercise 7.2

In questions 1 to 6, use the cosine formula to find the length of the remaining side.

1 $a = 17.1\,cm$, $c = 28.8\,cm$, $B = 108°$
2 $a = 7.86\,cm$, $b = 8.54\,cm$, $C = 37.42°$
3 $c = 17.5\,cm$, $b = 60.2\,cm$, $A = 63.67°$
4 $a = 18.5\,cm$, $b = 11.1\,cm$, $C = 120°$
5 $a = 4.31\,cm$, $b = 3.87\,cm$, $C = 29.23°$
6 $a = 7.59\,cm$, $c = 5.67\,cm$, $B = 72.23°$

In questions 7 to 10, find the angles of the triangle.

7 $a = 2\,cm$, $b = 3\,cm$, $c = 4\,cm$
8 $a = 5.4\,cm$, $b = 7.1\,cm$, $c = 8.3\,cm$
9 $a = 24\,cm$, $b = 19\,cm$, $c = 26\,cm$
10 $a = 2.60\,cm$, $b = 2.85\,cm$, $c = 4.70\,cm$
11 Find the largest angle of the triangle with sides 14 cm, 8.5 cm and 9 cm.
12 Find the smallest angle of the triangle with sides 6.4 cm, 5.7 cm and 8.2 cm.
13 Attempt to find the largest angle of a triangle with sides 5.8 cm, 8.3 cm and 14.1 cm. What happens and why?

In the remaining questions you may have to use either the sine formula or the cosine formula.

14 The shortest side of a triangle is 3.6 km long. Two of the angles are 37.25° and 48.4°. Find the length of the longest side.
15 The sides of a triangle are 123 m, 79 m and 97 m. Find its angles.
16 Given $b = 5.32\,cm$, $c = 6.47\,cm$, $A = 75.23°$, find B, C and a.
17 In a triangle ABC find the angle ACB when $c = 9.2\,cm$, $a = 5\,cm$ and $b = 11\,cm$.
18 The length of the side BC of a triangle ABC is 14.5 m, angle ABC = 71°, angle BAC = 57°. Calculate the lengths of the sides AC and AB.
19 In a quadrilateral $ABCD$, $AB = 3\,m$, $BC = 4\,m$, $CD = 7.4\,m$, $DA = 4.4\,m$ and the angle ABC is 90°. Determine the angle ADC.
20 Determine how many triangles exist with $a = 25\,cm$, $b = 30\,cm$, and $A = 50°$ and find the remaining sides and angles.
21 The length of the longest side of a triangle is 162 m. Two of the angles are 37.25° and 48.4°. Find the length of the shortest side.
22 In a quadrilateral $ABCD$, $AB = 4.3\,m$, $BC = 3.4\,m$ and $CD = 3.8\,m$. Angle ABC = 95° and angle BCD = 115°. Find the lengths of the diagonals.

23 From a point *O* on a straight line *OX*, lines *OP* and *OQ* of lengths 5 mm and 7 mm are drawn on the same side of *OX* so that angle *XOP* = 32° and angle *XOQ* = 55°. Find the length of *PQ*.

24 Two hooks *P* and *Q* on a horizontal beam are 30 cm apart. From *P* and *Q* strings *PR* and *QR*, 18 cm and 16 cm long respectively, support a weight at *R*. Find the distance of *R* from the beam and the angles which *PR* and *QR* make with the beam.

25 In a triangle *ABC*, *AB* is 5 cm long, angle *BAC* = 55° and angle *ABC* = 48°. Calculate the lengths of the sides *AC* and *BC* and the area of the triangle.

26 Two ships leave port at the same time. The first steams on a bearing 135° at 18 km h⁻¹ and the second on a bearing 205° at 15 km h⁻¹. Calculate the time that will have elapsed when they are 86 km apart.

27 *AB* is a base line of length 3 km, and *C*, *D* are points such that angle *BAC* = 32.25°, angle *ABC* = 119.08°, *DBC* = 60.17°, and angle *BCD* = 78.75°. The points *A* and *D* are on the same side of *BC*. Find the length of *CD*.

28 *ABCD* is a quadrilateral in which *AB* = 0.38 m, *BC* = 0.69 m, *AD* = 0.42 m, angle *ABC* = 109° and angle *BAD* = 123°. Find the area of the quadrilateral.

29 A weight was hung from a horizontal beam by two chains 8 m and 9 m long respectively, the ends of the chains being fastened to the same point of the weight, their other ends being fastened to the beam at points 10 m apart. Determine the angles which the chains make with the beam.

7.6 Introduction to surveying

The remainder of this chapter contains some examples of practical uses of the sine and cosine rules in surveying. Formulae for these uses are not given, because there may be a temptation to learn them; their importance is simply that they use the sine and cosine rules.

7.7 Finding the height of a distant object

Three forms of this problem are considered.

1 The point vertically beneath the top of the object is accessible

In Figure 7.8, A is the top of a tall object whose height h is required, and B is at its foot, on the same horizontal level as O. As B is accessible, you can measure the horizontal distance OB. Call this distance d. By using a theodolite you can find θ, the angle of elevation of AB.

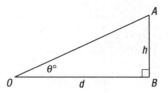

figure 7.8

Then
$$\tan\theta° = \frac{h}{d}$$

so
$$h = d\tan\theta°.$$

2 The point on the ground vertically beneath the top of the object is not accessible

In Figure 7.9 AB is the height to be found and B is not accessible. To find AB you can proceed as follows.

From a suitable point Q, use a theodolite to measure the angle $\theta°$.

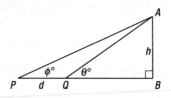

figure 7.9

Then measure a distance PQ, call it d, so that B, P and Q are in a straight line.

Then measure the angle $\phi°$.

- In triangle APQ, you can calculate the length AQ.
- Then, in triangle AQB, you can calculate h.

3 By measuring a horizontal distance in any direction

It may not be easy to obtain a distance PQ as in the previous example, where B, P and Q are in a straight line.

You can then use the following method.

In Figure 7.10 let AB be the height you need to find.

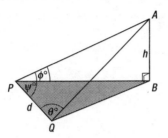

figure 7.10

Take a point P on the same level as B, and measure a horizontal distance PQ in any suitable direction. Let this distance be d.

At P measure angle $\phi°$, the angle of elevation of A, and angle ψ.

At Q measure angle $\theta°$.

- In triangle APQ you can use the sine formula to find AP.

- In triangle APB you can use the sine formula to calculate h.

You can also calculate the distances PB and QB if you need them.

7.8 Distance of an inaccessible object

Suppose you are at P and that at A is an inaccessible object whose distance from P you need to find. See Sigure 7.11.

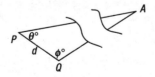

figure 7.11

Measure a distance PQ, call it d, in a convenient direction, and angle $\theta°$. Also measure angle $\phi°$.

In triangle APQ, you can use the sine formula to calculate AP.

7.9 Distance between two inaccessible but visible objects

Let A and B be two distant inaccessible objects at the same level. See Figure 7.12.

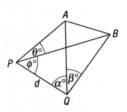

figure 7.12

Measure the length d of a convenient base line PQ at the same level as A and B.

At P measure angles $\theta°$ and $\phi°$, and at Q measure angles $\alpha°$ and $\beta°$.

In triangle APQ, you can use the sine formula to find AQ.

Similarly in triangle BPQ you can find QB.

Then in triangle AQB, you can use the cosine formula to find AB.

7.10 Triangulation

The methods employed in the last two examples are, in principle, those which are used in triangulation. This is the name given to the method used to survey a district, and to calculate its area. In practice, you need corrections to allow for sea level, and, over large areas, the curvature of the earth, but over small areas, the errors are small.

The method used is as follows.

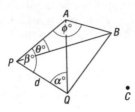

figure 7.13

Mark out and measure the distance PQ (Figure 7.13), called a **base line**, very accurately on suitable ground. Then select a point A and measure angles $\alpha°$ and $\beta°$.

You can then calculate the length AP.

Next select another point B and measure the angles $\theta°$ and $\phi°$, and calculate the length AB.

You now have enough information to calculate the area of the quadrilateral $PQBA$.

By joining BQ and by measuring the angles which BP and BQ make with PQ, you can calculate the area of quadrilateral $PQBA$ in a different way as a check on your results.

You can now select a new point C and continue by using the same methods.

By repeating this process with other points, you can create a network of triangles to cover a whole district.

As all the measurements of distance are calculated from the original distance d, it is essential that you measure the base line with minute accuracy. Similarly you need to measure the angles extremely accurately. You should build in checks at each stage, such as adding the angles of a triangle to see if their sum is 180°.

As a further check at the end of the work, or at any convenient stage, one of the lines whose length has been found by calculation, founded on previous calculations, can be used as a base line, and the whole survey worked backwards, finishing with the calculation of the original measured base line.

Example 7.8

Two points lie due west of a stationary balloon and are 1000 m apart. The angles of elevation of the balloon at the two points are 21.25° and 18°. Find the height of the balloon.

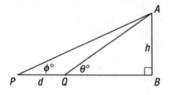

figure 7.14

In Figure 7.14, let A be the position of the balloon, and let its height be h metres.

$\theta = 21.25$, $\phi = 18$ and $d = 1000$ m.

Then, in triangle APQ, the third angle is 3.25.

In triangle APQ, using the sine formula,

$$\frac{\sin 18°}{AQ} = \frac{\sin 3.25°}{1000},$$

so
$$AQ = \frac{1000 \sin 18°}{\sin 3.25°}.$$

In triangle ABQ, $\qquad \sin 21.25° = \dfrac{h}{AQ}$

so $\quad h = AQ \sin 21.25° = \dfrac{1000 \sin 18° \sin 21.25°}{\sin 3.25°} = 1976.$

The height of the balloon is approximately 1976 metres. ∎

Example 7.9

A balloon is observed from two stations A and B at the same horizontal level, A being 1000 m north of B. At a given instant the bearing of the balloon from A is 033.2° and its angle of elevation is 53.42°, while from B its bearing is 021.45°. Calculate the height of the balloon.

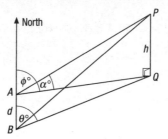

figure 7.15

In Figure 7.15, the balloon is at P, and Q is directly below P.

ϕ = 33.2, θ = 21.45, α = 53.42 and d = 1000 m.

Using the sine formula in triangle ABQ to find AQ, noting that angle $BQA = \phi° - \theta° = 33.2° - 21.45° = 11.75°$,

$$\frac{\sin 21.45°}{AQ} = \frac{\sin 11.75°}{1000}$$

so

$$AQ = \frac{1000 \sin 21.45°}{\sin 11.75°}.$$

In triangle APQ $\tan 53.42° = \dfrac{h}{AQ}$

so

$$h = AQ \tan 53.42° = \frac{1000 \sin 21.45° \sin 53.42°}{\sin 11.75°} = 2419.7.$$

The height of the balloon is therefore 2420 metres approximately. ∎

Example 7.10

A surveyor who wishes to find the width of a river measures along a level stretch on one bank a line AB, 150 m long.

From A the surveyor observes that a post P on the opposite bank is placed so that angle $PAB = 51.33°$, and angle $PBA = 69.20°$. What was the width of the river?

In Figure 7.16, AB is the measured distance, 150 m. P is the post on the other side of the river. PQ, which is drawn perpendicular to AB, is the width w of the river. The angles α and β are 51.33° and 69.20°.

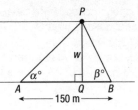

figure 7.16

To find PQ, first calculate AP from triangle APB.

Then
$$\frac{\sin 69.20°}{AP} = \frac{\sin (180 - 51.33 - 69.20)°}{150}$$

so
$$AP = \frac{150 \sin 69.20°}{\sin 59.47°}.$$

Then
$$\sin 51.33° = \frac{w}{AP}$$

so $w = AP \sin 51.33° = \dfrac{150 \sin 69.20° \sin 51.33°}{\sin 59.47°} = 127.1.$

Therefore the width of the river is 127 metres approximately. ∎

Exercise 7.3

1 A surveyor, who measures the angle of elevation of a tree as 32° and then walks 8 m directly towards the tree, finds that the new angle of elevation is 43°. Calculate the height of the tree.

2 From a point Q on a horizontal plane the angle of elevation of the top of a distant mountain is 22.3°. At a point P, 500 m further away in a direct horizontal line, the angle of elevation of the mountain is 16.6°. Find the height of the mountain.

3 Two people, 1.5 km apart, stand on opposite sides of a church steeple and in the same straight line with it. From one, the angle of elevation of the top of the tower is 15.5° and, from the other, 28.67°. Calculate the height of the steeple in metres.

4 A surveyor, who wishes to find the width of a river, stands on one bank of the river and measures the angle of elevation of a high building on the edge of the other bank and directly opposite as 31°. After walking 110 m away from the river in the straight line from the building the surveyor finds that the angle of elevation of the building is now 20.92°. Calculate the width of the river.

5 *A* and *B* are two points on opposite sides of swampy ground. From a point *P* outside the swamp it is found that *PA* is 882 metres and *PB* is 1008 metres. The angle subtended at *P* by *AB* is 55.67°. Calculate the distance between *A* and *B*.

6 *A* and *B* are two points 1.8 km apart on a level piece of ground along the bank of a river. *P* is a post on the opposite bank. It is found that angle *PAB* = 62° and angle *PBA* = 48°. Calculate the width of the river.

7 The angle of elevation of the top of a mountain from the bottom of a tower 180 m high is 26.42°. From the top of the tower the angle of elevation is 25.3°. Calculate the height of the mountain.

8 Two observers 5 km apart measure the bearing of the base of the balloon and the angle of elevation of the balloon at the same instant. One finds that the bearing is 041°, and the elevation is 24°. The other finds that the bearing is 032°, and the elevation is 26.62°. Calculate the height of the balloon.

9 Two landmarks *A* and *B* are observed from a point *P* to be in a line due east. From a point *Q* 4.5 km in a direction 060° from *P*, *A* is observed to be due south while *B* is on a bearing 128°. Find the distance between *A* and *B*.

10 At a point *P* in a straight road *PQ* it is observed that two distant objects *A* and *B* are in a straight line making an angle of 35° at *P* with *PQ*. At a point *C* 2 km along the road from *P* it is observed that angle *ACP* is 50° and angle *BCQ* is 64°. Calculate the distance between *A* and *B*.

11 An object *P* is situated 345 m above a level plane. Two people, *A* and *B*, are standing on the plane, *A* in a direction south-west of *P* and *B* due south of *P*. The angles of elevation of *P* as observed at *A* and *B* are 34° and 26° respectively. Find the distance between *A* and *B*.

12 *P* and *Q* are points on a straight coast line, *Q* being 5.3 km east of *P*. A ship starting from *P* steams 4 km in a direction 024.5°. Calculate: (a) the distance the ship is now from the coast-line; (b) the ship's bearing from *Q*; (c) the distance of the ship from *Q*.

13 At a point *A* due south of a chimney stack, the angle of elevation of the stack is 55°. From *B*, due west of *A*, such that *AB* = 100 m, the elevation of the stack is 33°. Find the height of the stack and its horizontal distance from *A*.

14 *AB* is a base line 0.5 km long and *B* is due west of *A*. At *B* a point *P* has bearing 335.7°. The bearing of *P* from *A* is 314.25°. How far is *P* from *A*?

15 A horizontal bridge over a river is 380 m long. From one end, *A*, it is observed that the angle of depression of an object, *P*, on the surface of the water vertically beneath the bridge, is 34°. From the other end, *B*, the angle of depression of the object is 62°. What is the height of the bridge above the water?

16 A straight line *AB*, 115 m long, lies in the same horizontal plane as the foot *Q* of a church tower *PQ*. The angle of elevation of the top of the tower at *A* is 35°. Angle *QAB* is 62° and angle *QBA* is 48°. What is the height of the tower?

17 *A* and *B* are two points 1500 metres apart on a road running due west. A soldier at *A* observes that the bearing of an enemy's battery is 295.8°, and at *B*, 301.5°. The range of the guns in the battery is 5 km. How far can the soldier go along the road from *A* before being within range and what length of the road is within range?

08

radians

In this chapter you will learn:
- that you can measure an angle in radians as an alternative to degrees
- the formulae for length of a circular arc and the area of a circular sector
- how to convert from radians to degrees and vice versa.

8.1 Introduction

Who decided that there should be 360 degrees in a full circle, and therefore 90 degrees in a right angle? It is not the answer to this question which is important – it was actually the Babylonians – it is the fact that the question exists at all. Someone, somewhere did make the decision that the unit for angle should be the degree as we now know it. However, it could just have equally been 80 divisions which make a right angle or 100 divisions. So it seems worth asking, is there a best unit for measuring angle? Or is there a better choice for this unit than the degree?

It turns out that the answer is yes. A better unit is the radian.

8.2 Radians

A **radian** is the angle subtended at the centre of a circle by a circular arc equal in length to the radius. See Figure 8.1.

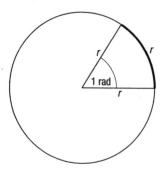

figure 8.1

The angle of 1 radian is written as 1 rad, but if no units are given for angles you should assume that the unit is radians.

If you are using radians with a calculator, you will need to make sure that the calculator is in radian mode. If necessary, look up how to use radian mode in the manual.

8.3 Length of a circular arc

The right-hand diagram in Figure 8.2 shows a circle of radius r cm with an angle of θ rad at the centre. The left-hand diagram shows a circle with the same radius but with an angle of 1 rad at the centre.

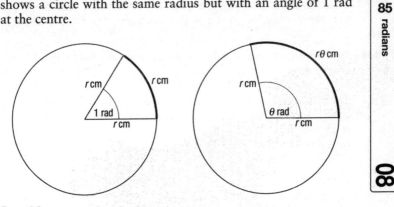

figure 8.2

Look at the relationship between the left- and right-hand diagrams. As the angle at the centre of the circle in the left-hand diagram has been multiplied by a factor θ to get the right-hand diagram, so has the arc length. The new arc length is therefore $\theta \times$ the original arc length r cm and therefore $r\theta$ cm.

If you call the arc length s cm, then

$$s = r\theta.$$ **Equation 1**

Example 8.1

O is the centre of a circle of radius 3 cm. The points A and B lie on its circumference and angle $AOB = 2$ rad. Find the length of the perimeter of the segment bounded by the arc AB and the chord AB.

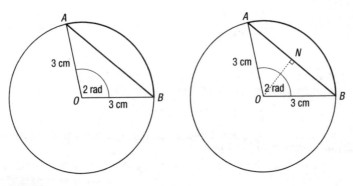

figure 8.3

The left-hand diagram in Figure 8.3 shows this situation. The perimeter must be found in two sections, the arc AB and the chord AB.

The length of the arc AB is given by using Equation 1,

$$\text{length of arc } AB = 3 \times 2 = 6\,\text{cm}.$$

To find the length of the chord AB, drop the perpendicular from O to AB meeting AB at N, shown in the right-hand part of Figure 8.3.

Then $\qquad AB = 2AN = 2 \times (3\sin 1) = 5.05.$

The perimeter is then given by

$$\text{perimeter} = \text{arc } AB + AB = (6 + 5.05)\,\text{cm} = 11.05\,\text{cm}. \ \blacksquare$$

8.4 Converting from radians to degrees

Consider the case when the arc of a circle of radius r cm is actually the complete circumference of the circle. In this case, the arc length is $2\pi r$ cm.

Suppose that the angle at the centre of this arc is θ rad. Then, using Equation 1, the length of the arc is $r\theta$ cm.

Then it follows that $2\pi r = r\theta$, so that the angle at the centre is 2π rad.

But as the angle at the centre of the circle is $360°$,

$$2\pi \text{ rad} = 360°.$$

Therefore $\qquad \pi \text{ rad} = 180°.$

This equation, π rad $= 180°$, is the one you should remember when you need to change from degrees to radians, and vice versa.

In many cases, when angles such as $45°$ and $60°$ are given in radians they are given as multiples of π. That is

$$\tfrac{1}{4}\pi \text{ rad} = 45° \text{ and } \tfrac{1}{3}\pi \text{ rad} = 60°.$$

You can also work out 1 radian in degrees from the equation π rad $= 180°$. You find that

$$1 \text{ rad} = 57.296°.$$

This equation is very rarely used in practice. When you need to convert radians to degrees or vice versa, use the fact that π rad = 180° and use either $\frac{180}{\pi}$ or $\frac{\pi}{180}$ as a conversion factor.

Thus, for example,

$$10° = 10 \times \frac{\pi}{180} \text{ rad} = \tfrac{1}{18} \pi \text{ rad}.$$

8.5 Area of a circular sector

Figure 8.4 shows a shaded sector of a circle with radius r units and an angle at the centre of θ rad. Let the area of the shaded region be A units².

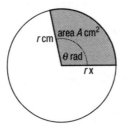

figure 8.4

The area A units² is a fraction of the area of the whole circle. As the area of the whole circle is πr^2 units², and the angle at the centre is 2π rad, when the angle at the centre is θ rad the shaded area is a fraction $\frac{\theta}{2\pi}$ of the total area of the circle.

Therefore the area, in units², of the circular arc is

$$\frac{\theta}{2\pi} \times \pi r^2 = \tfrac{1}{2}r^2\theta. \qquad \textbf{Equation 2}$$

Example 8.2
O is the centre of a circle of radius 3 cm. The points A and B lie on its circumference and angle $AOB = 2$ rad. Find the area of the segment bounded by the arc AB and the chord AB.

Figure 8.3 shows this situation. The area of the segment must be found by finding the area of the whole sector OAB, and then subtracting the area of the triangle OAB.

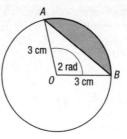

figure 8.5

The area A cm^2 of the sector OAB is given by using Equation 2,

$$A = \tfrac{1}{2}r^2\theta = \tfrac{1}{2} \times 3^2 \times 2 = 9.$$

To find the area of triangle OAB, use the formula $\tfrac{1}{2}ab \sin C$, given on page 65.

Then

area of triangle OAB = $\tfrac{1}{2}ab \sin C = \tfrac{1}{2} \times 3^2 \times \sin 2 = 4.092.$

The area of the shaded segment is then given by

area of segment = area of sector OAB – area of triangle OAB

$$= (9 - 4.092) \text{ cm}^2 = 4.908 \text{ cm}^2. \blacksquare$$

Exercise 8

In questions 1 to 6, write down the number of degrees in each of the angles which are given in radians.

1 $\frac{1}{3}\pi$ **2** $\frac{1}{12}\pi$

3 $\frac{3}{2}\pi$ **4** $\frac{2}{3}\pi$

5 $\frac{3}{4}\pi$ **6** 4π

In questions 7 to 12, find the values of the given ratios.

7 $\sin \frac{1}{5}\pi$ **8** $\cos \frac{1}{8}\pi$

9 $\sin \frac{1}{10}\pi$ **10** $\cos \frac{3}{8}\pi$

11 $\sin(\frac{1}{3}\pi + \frac{1}{4}\pi)$ **12** $\sin \frac{1}{6}\pi$

13 Give the angle 0.234 rad in degrees correct to two decimal places.

In questions 14 to 17, express the following angles in radians, using fractions of π.

14 15°   **15** 72°

16 66°   **17** 105°

18 Find in radians the angle subtended at the centre of a circle of radius 2.4 cm by a circular arc of length 11.4 cm.

19 Find the length of the circular arc which subtends an angle of 0.31 rad at the centre of a circle of radius 3.6 cm.

20 Find the area of the circular sector which subtends an arc of 2.54 rad at the centre of a circle of radius 2.3 cm.

21 Find in radians the angle that a circular sector of area 20 cm² subtends at the centre of a circle of radius 5 cm.

22 A circular arc is 154 cm long and the radius of the arc is 252 cm. Find the angle subtended at the centre of the circle, in radians and degrees.

23 The angles of a triangle are in the ratio of 3 : 4 : 5. Express them in radians.

24 A chord of length 8 cm divides a circle of radius 5 cm into two parts. Find the area of each part.

25 Two circles each of radius 4 cm overlap, and the length of their common chord is also 4 cm. Find the area of the overlapping region.

26 A new five-sided coin is to be made in the shape of figure 8.6.

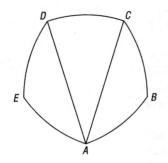

figure 8.6

The point *A* on the circumference of the coin is the centre of the arc *CD*, which has a radius of 2 cm. Similarly *B* is the centre of the arc *DE*, and so on. Find the area of one face of the coin.

09

relations between the ratios

In this chapter you will learn:
- some relations between the sine and cosine of an angle
- the trigonometric form of Pythagoras's theorem
- the meaning of secant, cosecant and cotangent.

9.1 Introduction

In Section 3.2, pages 20 and 21, you saw that for any angle $\theta°$:

$$\sin(90 - \theta)° = \cos\theta°,$$

$$\cos(90 - \theta)° = \sin\theta°,$$

$$\sin^2\theta + \cos^2\theta = 1.$$

The third of these relations is a form of Pythagoras's theorem, and it sometimes goes by that name.

In Section 5.6, page 52, you saw that

$$\tan\theta = \frac{\sin\theta}{\cos\theta}.$$

In Sections 6.2 to 6.4, pages 55, 58 and 59, you saw that

$$\sin\theta° = \sin(180 - \theta)°,$$

$$\cos\theta° = \cos(-\theta)°,$$

$$\tan\theta° = \tan(180 + \theta)°.$$

In this chapter, you will explore these and other relations, as well as meeting the new ratios secant, cosecant and cotangent.

You will also learn to solve a wider variety of trigonometric equations, using these rules to help.

Some of the relations given above hold whatever units are used for measuring angles. Examples are $\sin^2\theta + \cos^2\theta = 1$ and $\frac{\sin\theta}{\cos\theta} = \tan\theta$. When this is the case no units for angle are given. However, for some of the relations the angles must be measured in degrees for the relation to be true. This is the case for $\sin(90 - \theta)° = \cos\theta°$ and $\sin(180 - \theta)° = \sin\theta°$. In these cases, degree signs will be used.

9.2 Secant, cosecant and cotangent

The three relations secant, cosecant and cotangent, usually abbreviated to sec, cosec and cot, are defined by the rules

$$\sec\theta = \frac{1}{\cos\theta},$$

$$\text{cosec}\,\theta = \frac{1}{\sin\theta},$$

$$\cot \theta = \frac{1}{\tan \theta},$$

provided $\cos \theta$, $\sin \theta$ and $\tan \theta$ are not zero.

In the early part of the twentieth century, tables were used to find values of the trigonometric ratios. There used to be tables for sec, cosec and cot, but these have now all but disappeared, and if you want their values from a calculator, you need to use the definitions above.

You can write Pythagoras's theorem, $\sin^2 \theta + \cos^2 \theta = 1$, in terms of these new ratios.

Divide every term of $\sin^2 \theta + \cos^2 \theta = 1$ by $\cos^2 \theta$ to obtain

$$\frac{\sin^2 \theta}{\cos^2 \theta} + \frac{\cos^2 \theta}{\cos^2 \theta} = \frac{1}{\cos^2 \theta},$$

which simplifies to $\quad \tan^2 \theta + 1 = \sec^2 \theta$.

Similarly, by dividing $\sin^2 \theta + \cos^2 \theta = 1$ by $\sin^2 \theta$ you can show that

$$1 + \cot^2 \theta = \operatorname{cosec}^2 \theta.$$

Example 9.1

Let $\cos x = \frac{3}{5}$. Find the possible values of $\tan x$, $\sec x$ and $\operatorname{cosec} x$.

Using Pythagoras's theorem, $\sin^2 x + \cos^2 x = 1$, the value of $\sin x$ is

$$\sin x = \pm \sqrt{1 - \left(\tfrac{3}{5}\right)^2} = \pm \tfrac{4}{5}.$$

Then, using $\tan x = \dfrac{\sin x}{\cos x}$,

$$\tan x = \frac{\sin x}{\cos x} = \frac{\pm \frac{4}{5}}{\frac{3}{5}} = \pm \tfrac{4}{3}.$$

As $\sec x = \dfrac{1}{\cos x}$, $\sec x = \dfrac{1}{\frac{3}{5}} = \tfrac{5}{3}$.

As $\operatorname{cosec} x = \dfrac{1}{\sin x}$, $\operatorname{cosec} x = \dfrac{1}{\pm \frac{4}{5}} = \pm \tfrac{5}{4}$. ∎

Example 9.2

Solve the equation $3 \cos^2 \theta° = 1 - 2 \sin \theta°$ giving solutions in the interval −180 to 180.

If you substitute $\cos^2\theta° = 1 - \sin^2\theta°$ you obtain an equation in which every term, except the constant, is a multiple of a power of $\sin\theta$, that is, a polynomial equation in $\sin\theta°$. You can solve this by the usual methods.

Thus
$$3\cos^2\theta° = 1 - 2\sin\theta°$$
$$3(1 - \sin^2\theta°) = 1 - 2\sin\theta°$$
$$3\sin^2\theta° - 2\sin\theta° - 2 = 0.$$

This is a quadratic equation in $\sin\theta$. Using the quadratic equation formula,

$$\sin\theta° = \frac{2 \pm \sqrt{28}}{6},$$

so
$$\sin\theta° = 1.215... \text{ or } \sin\theta° = -0.5485... .$$

The first of these is impossible. The principal angle corresponding to the second is −33.27.

Then $180 - (-33.27) = 213.27$ is also a solution. (See page 55.) But this is outside the required range, so subtract 360 to get

$$213.27 - 360 = -146.73.$$

Thus the solutions are −33.27 and −146.73. ∎

Example 9.3
Solve the equation $\cos\theta° = 1 + \sec\theta°$ giving all the solutions in the interval from 0 to 360.

Notice that if you write $\sec\theta° = \dfrac{1}{\cos\theta°}$ all the terms in the equation will involve $\cos\theta°$. Therefore

$$\cos\theta° = 1 + \frac{1}{\cos\theta°}$$
$$\cos^2\theta° = \cos\theta° + 1$$
$$\cos^2\theta° - \cos\theta° - 1 = 0$$
$$\cos\theta° = \frac{1 \pm \sqrt{5}}{2} = 1.618... \text{ or } -0.618... .$$

The first of these solutions is impossible. The principal angle corresponding to the second is 128.17.

Then $360 - 128.17 = 231.83$ is also a solution. (See page 59.)

Thus the solutions are 128.17 and 308.17. ∎

Example 9.4

Solve the equation $2 \sec \theta° = 2 + \tan^2 \theta°$ giving all the solutions in the interval from -180 to 180.

If you use Pythagoras's theorem in the form $\tan^2 \theta° = \sec^2 \theta° - 1$ to substitute for $\tan^2 \theta°$ all the terms in the equation will involve $\sec \theta°$.

Therefore
$$2 \sec \theta° = 2 + \tan^2 \theta°$$
$$= 2 + (\sec^2 \theta° - 1)$$
$$\sec^2 \theta° - 2 \sec \theta° + 1 = 0$$
$$(\sec \theta° - 1)^2 = 0$$
$$\sec \theta° = 1.$$

Therefore
$$\cos \theta° = \frac{1}{\sec \theta°} = 1$$

so
$$\theta = 0.$$

Thus the solution is 0. ∎

Example 9.5

Solve the equation $\tan \theta = 2 \sin \theta$ giving all solutions between $-\pi$ and π inclusive.

It is often useful to write equations in terms of the sine and cosine functions, because there are so many more simplifying equations which you can use.

So
$$\tan \theta = 2 \sin \theta$$
$$\frac{\sin \theta}{\cos \theta} = 2 \sin \theta.$$

Multiplying both sides of this equation by $\cos \theta$ gives
$$\sin \theta - 2 \sin \theta \cos \theta = 0.$$

Factorizing,
$$\sin \theta (1 - 2 \cos \theta) = 0$$

so
$$\sin \theta = 0 \text{ or } \cos \theta = \tfrac{1}{2}.$$

Using the methods of Sections 6.2 and 6.3, solve these two equations for θ.

When $\sin \theta = 0$, $\theta = -\pi$ or 0 or π; when $\cos \theta = \tfrac{1}{2}$, $\theta = -\tfrac{1}{3}\pi$ or $\tfrac{1}{3}\pi$.

Therefore $\theta = -\pi, -\tfrac{1}{3}\pi, 0, \tfrac{1}{3}\pi$ or π. ∎

Exercise 9

1 Find the value of $\cos\theta$ given that $\sin\theta = 0.8192$, and that θ is obtuse.

2 Find the possible values of $\tan\theta$ given that $\cos\theta = 0.3$.

3 Find the possible values of $\sec\theta$ when $\tan\theta = 0.4$.

4 The angle α is acute, and $\sec\alpha = k$. Find in terms of k the value of $\csc\alpha$.

5 Let $\tan\theta° = t$, where θ lies between 90 and 180. Calculate, in terms of t, the values of $\sec\theta°$ $\cos\theta°$ and $\sin\theta°$.

6 Let $\sec\theta = s$, where θ is acute. Find the values of $\cot\theta$ and $\sin\theta$ in terms of s.

In questions 7 to 12, solve the given equation for θ, giving your answers in the interval from −180 to 180.

7 $\cos^2\theta° = \frac{1}{4}$

8 $2\sin\theta° = \csc\theta°$

9 $2\sin^2\theta° - \sin\theta° = 0$

10 $2\cos^2\theta° = 3\sin\theta° + 2$

11 $\tan\theta° = \cos\theta°$

12 $\sin\theta° = 2\cos\theta°$

13 $2\sec\theta° = \csc\theta°$

14 $5(1 - \cos\theta°) = 4\sin^2\theta°$

15 $4\sin\theta° \cos\theta° + 1 = 2(\sin\theta° + \cos\theta°)$

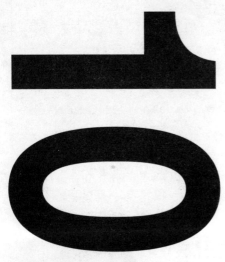

10

ratios of compound angles

In this chapter you will learn:

- how to find the values of $\sin(A + B)$, $\cos(A + B)$ and $\tan(A + B)$ knowing the values of the sine, cosine and tangent of A and B
- how to modify these formulae for $\sin(A - B)$, $\cos(A - B)$ and $\tan(A - B)$
- how to find the values of $\sin 2A$, $\cos 2A$ and $\tan 2A$ knowing the values of the sine, cosine and tangent of A.

10.1 Compound angles

A compound angle is an angle of the form $A + B$ or $A - B$. This chapter is about finding the sine, cosine and tangent of $A + B$ or $A - B$ in terms of the sine, cosine and tangent, as appropriate, of the individual angles A and B.

Notice immediately that $\sin(A + B)$ is not equal to $\sin A + \sin B$. You can try this for various angles, but if it were true, then

$$\sin 180° = \sin 90° + \sin 90°$$
$$= 1 + 1 = 2,$$

which is clearly false.

It is difficult to give general proofs of formulae for $\sin(A + B)$ and $\cos(A + B)$, and this is not attempted in this book. Proofs which apply only to angles in a restricted range are given. The formulae obtained will then be assumed to be true for all angles.

10.2 Formulae for sin(*A* + *B*) and sin(*A* − *B*)

Suppose that angles A and B are both between 0 and 90°. In Figure 10.1, the angles A and B are drawn at the point Q, and the line QN is drawn of length h. PR is perpendicular to QN, and meets QP and QR at P and R respectively.

Let the lengths of PQ, QR, PN and NR be r, p, x and y respectively, as shown in the diagram.

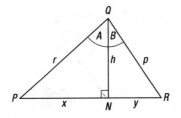

figure 10.1

The strategy for deriving the formula for $\sin(A + B)$ is to say that the area of triangle PQR is the sum of the areas of triangles PQN and RQN.

Since the formula for area of a triangle is $\frac{1}{2}ab \sin C$, the area of triangle PQR is $\frac{1}{2}rp \sin(A + B)$ and of triangles PQN and RQN are $\frac{1}{2}rh \sin A$ and $\frac{1}{2}ph \sin B$ respectively.

Then area of triangle PQR = area of triangle PQN + area of triangle RQN so

$$\frac{1}{2}rp \sin(A + B) = \frac{1}{2}rh \sin A + \frac{1}{2}ph \sin B.$$

Then, multiplying both sides of the equation by 2, and dividing both sides by rp gives

$$\sin(A + B) = \frac{h}{p} \sin A + \frac{h}{r} \sin B.$$

Noticing that $\frac{h}{p} = \cos B$ and $\frac{h}{r} = \cos A$, the formula becomes

$$\sin(A + B) = \cos B \times \sin A + \cos A \times \sin B.$$

This equation is usually written as

$$\sin(A + B) = \sin A \cos B + \cos A \sin B. \quad \textbf{Equation 1}$$

Although Equation 1 has been proved only for angles A and B which are acute, the result is actually true for all angles A and B, positive and negative. From now on you may assume this result.

You can use a similar method based on the difference of two areas to derive a formula for $\sin(A - B)$ from Figure 10.2. (You are asked to derive this formula in Exercise 10.1, question 15.)

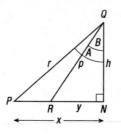

figure 10.2

You would then get the formula

$$\sin(A - B) = \sin A \cos B - \cos A \sin B. \quad \textbf{Equation 2}$$

10.3 Formulae for cos(A + B) and cos(A − B)

Suppose that angles A and B are both between 0 and 90°. In Figure 10.3, which is the same as Figure 10.1, the angles A and B are drawn at the point Q, and the line QN is drawn of length h. PR is perpendicular to QN, and meets QP and QR at P and R respectively.

Let the lengths of PQ, QR, PN and NR be r, p, x and y respectively, as shown in the diagram.

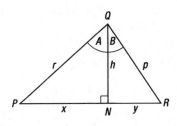

figure 10.3

The strategy in this case is to use the cosine formula to derive an expression for $\cos(A + B)$.

In triangle PQR

$$(x + y)^2 = r^2 + p^2 - 2rp\cos(A + B),$$

so, simplifying, and using Pythagoras's theorem

$$
\begin{aligned}
2rp\cos(A + B) &= r^2 + p^2 - (x + y)^2 \\
&= r^2 + p^2 - x^2 - 2xy - y^2 \\
&= (r^2 - x^2) + (p^2 - y^2) - 2xy \\
&= h^2 + h^2 - 2xy = 2h^2 - 2xy.
\end{aligned}
$$

After dividing both sides by $2rp$ you obtain

$$\cos(A + B) = \frac{h^2}{rp} - \frac{xy}{rp}.$$

Noticing that $\frac{h}{r} = \cos A$, $\frac{h}{p} = \cos B$, $\frac{x}{r} = \sin A$ and $\frac{y}{p} = \sin B$, the formula becomes

$$\cos(A + B) = \cos A \cos B - \sin A \sin B. \quad \textbf{Equation 3}$$

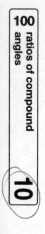

You can use a similar method based on Figure 10.2 to derive a formula for $\cos(A - B)$. (You are asked to derive this formula in Exercise 10.1, question 16.)

You would then get the formula

$$\cos(A - B) = \cos A \cos B + \sin A \sin B. \quad \textbf{Equation 4}$$

10.4 Formulae for tan(A + B) and tan(A − B)

You can use Equations 1 and 3 to derive a formula for $\tan(A + B)$, starting from the formula

$$\tan \theta = \frac{\sin \theta}{\cos \theta}.$$

Then

$$\tan(A + B) = \frac{\sin(A + B)}{\cos(A + B)}$$

$$= \frac{\sin A \cos B + \cos A \sin B}{\cos A \cos B - \sin A \sin B}.$$

Now divide the numerator and denominator of this fraction by $\cos A \cos B$. Then

$$\tan(A + B) = \frac{\sin A \cos B + \cos A \sin B}{\cos A \cos B - \sin A \sin B}$$

$$= \frac{\dfrac{\sin A \cos B}{\cos A \cos B} + \dfrac{\cos A \sin B}{\cos A \cos B}}{\dfrac{\cos A \cos B}{\cos A \cos B} - \dfrac{\sin A \sin B}{\cos A \cos B}}$$

$$= \frac{\tan A + \tan B}{1 - \tan A \tan B}.$$

Therefore

$$\tan(A + B) = \frac{\tan A + \tan B}{1 - \tan A \tan B}. \quad \textbf{Equation 5}$$

As with the formulae for $\sin(A - B)$ and $\cos(A - B)$, you can use a similar method to derive a formula for $\tan(A - B)$. (You are asked to derive this formula in Exercise 10.1, question 17.)

You would then get the formula

$$\tan(A - B) = \frac{\tan A - \tan B}{1 + \tan A \tan B}. \quad \textbf{Equation 6}$$

10.5 Worked examples

Example 10.1

Using the values of the sines and cosines of 30° and 45° in Section 3.4, page 26, find the exact values of $\sin 75°$ and $\cos 15°$.

Using $\qquad \sin(A + B) = \sin A \cos B + \cos A \sin B$

and substituting $\qquad A = 45°$ and $B = 30°$,

you find $\qquad \sin 75° = \sin 45° \cos 30° + \cos 45° \sin 30°$

$$= \frac{\sqrt{2}}{2} \times \frac{\sqrt{3}}{2} + \frac{\sqrt{2}}{2} \times \frac{1}{2}$$

$$= \frac{\sqrt{6} + \sqrt{2}}{4}.$$

To find $\cos 15°$, you could either note that $\cos \theta° = \sin(90 - \theta)°$, and therefore $\cos 15° = \sin 75°$, or you could say that

$$\cos 15° = \cos(45° - 30°)$$

$$= \cos 45° \cos 30° + \sin 45° \sin 30°$$

$$= \frac{\sqrt{2}}{2} \times \frac{\sqrt{3}}{2} + \frac{\sqrt{2}}{2} \times \frac{1}{2}$$

$$= \frac{\sqrt{6} + \sqrt{2}}{4}. \quad \blacksquare$$

Example 10.2

Let the angles α and β be acute, and such that $\cos \alpha = 0.6$ and $\cos \beta = 0.8$. Calculate the exact values of $\sin(\alpha + \beta)$ and $\cos(\alpha + \beta)$.

First you need the values of $\sin \alpha$ and $\sin \beta$. You can do this by using Pythagoras's theorem in the form

$$\sin^2 \theta + \cos^2 \theta = 1.$$

Then $\qquad \sin^2 \alpha = 1 - \cos^2 \alpha$

$$= 1 - 0.36 = 0.64.$$

As α is acute, $\sin \alpha$ is positive, so $\sin \alpha = 0.8$.

Similarly, as β is acute,

$$\sin\beta = \sqrt{1 - \cos^2\beta}$$
$$= \sqrt{1 - 0.64} = \sqrt{0.36}$$
$$= 0.6.$$

Then, using the formula $\sin(A + B) = \sin A \cos B + \cos A \sin B$,

$$\sin(\alpha + \beta) = 0.8 \times 0.8 + 0.6 \times 0.6$$
$$= 0.64 + 0.36$$
$$= 1.$$

Similarly

$$\cos(\alpha + \beta) = \cos\alpha \cos\beta - \sin\alpha \sin\beta$$
$$= 0.6 \times 0.8 - 0.8 \times 0.6$$
$$= 0.48 - 0.48$$
$$= 0.$$

So $\sin(\alpha + \beta) = 1$ and $\cos(\alpha + \beta) = 0$. ∎

Example 10.3

Use the formula for $\cos(A - B)$ to show that $\cos(270 - \theta)° = -\sin\theta°$.

Put $A = 270$ and $B = \theta$.

Then $\cos(270 - \theta)° = \cos 270° \cos\theta° + \sin 270° \sin\theta°$
$$= 0 \times \cos\theta° + (-1) \times \sin\theta°$$
$$= -\sin\theta°. ∎$$

Exercise 10.1

1 If $\cos A = 0.2$ and $\cos B = 0.5$, and angles A and B are acute, find the values of $\sin(A + B)$ and $\cos(A + B)$.

2 Use the exact values of sine and cosine of $30°$ and $45°$ to find the exact values of $\sin 15°$ and $\cos 75°$.

3 Use the formula for $\sin(A - B)$ to show that
$$\sin(90 - \theta)° = \cos\theta°.$$

4 Calculate the value of $\sin(A - B)$ when $\cos A = 0.309$ and $\sin B = 0.23$, given that angle A is acute and angle B is obtuse.

5 Let $\sin A = 0.71$ and $\cos B = 0.32$ where neither A nor B is a first quadrant angle. Find $\sin(A + B)$ and $\tan(A + B)$.

6 Use the formula for $\tan(A + B)$ to find the exact value, in terms of $\sqrt{2}$ and $\sqrt{3}$, of $\tan 75°$.

7 Find $\tan(A + B)$ and $\tan(A - B)$ given that $\tan A = 1.2$ and $\tan B = 0.4$.

8 By using the formula for $\tan(A - B)$, prove that
$$\tan(180 - \theta)° = -\tan\theta°.$$

9 Find the value of $\sin 52° \cos 18° - \cos 52° \sin 18°$.

10 Find the value of $\cos 73° \cos 12° + \sin 73° \sin 12°$.

11 Find the value of $\dfrac{\tan 52° + \tan 16°}{1 - \tan 52° \tan 16°}$.

12 Find the value of $\dfrac{\tan 64° - \tan 25°}{1 + \tan 64° \tan 25°}$.

13 Prove that $\sin(\theta + 45)° = \dfrac{1}{\sqrt{2}}(\sin\theta° + \cos\theta°)$.

14 Prove that $\tan(\theta + 45)° = \dfrac{1 + \tan\theta°}{1 - \tan\theta°}$.

15 Use the method of Section 10.2 to prove that
$\sin(A - B) = \sin A \cos B - \cos A \sin B$.

16 Use the method of Section 10.3 to prove that
$\cos(A - B) = \cos A \cos B + \sin A \sin B$.

17 Use the method of Section 10.4 to prove that
$$\tan(A - B) = \frac{\tan A - \tan B}{1 + \tan A - \tan B}.$$

10.6 Multiple angle formulae

From Equations 1, 3 and 5 you can deduce other important formulae.

In the formula $\sin(A + B) = \sin A \cos B + \cos A \sin B$, put $B = A$.

Then $\qquad \sin(A + A) = \sin A \cos A + \cos A \sin A$

so $\qquad \sin 2A = 2 \sin A \cos A.$ **Equation 7**

You may sometimes need to use this formula with $2A$ replaced by θ.

Then you obtain

$$\sin\theta = 2 \sin\tfrac{1}{2}\theta \cos\tfrac{1}{2}\theta. \qquad \textbf{Equation 8}$$

Equations 7 and 8 are really the same formula. Use whichever form is more convenient for the problem in hand.

Again, if you put $B = A$ in the formula

$$\cos(A + B) = \cos A \cos B - \sin A \sin B,$$

you get $\qquad \cos(A + A) = \cos A \cos A - \sin A \sin A,$

which simplifies to

$$\cos 2A = \cos^2 A - \sin^2 A. \qquad \text{Equation 9}$$

Note the way of writing $\cos A \times \cos A$ or $(\cos A)^2$ as $\cos^2 A$. This is used for positive powers, but is not usually used for writing powers such as $(\cos A)^{-1}$ because the notation $\cos^{-1} x$ is reserved for the angle whose cosine is x.

You can put Equation 9 into other forms using Pythagoras's theorem, $\sin^2 A + \cos^2 A = 1$. Writing $\sin^2 A = 1 - \cos^2 A$ in Equation 9 you obtain

$$\cos 2A = \cos^2 A - \sin^2 A$$
$$= \cos^2 A - (1 - \cos^2 A)$$
$$= 2 \cos^2 A - 1$$

so $\qquad \cos 2A = 2 \cos^2 A - 1. \qquad \text{Equation 10}$

On the other hand, if you put $\cos^2 A = 1 - \sin^2 A$ in Equation 9 you get

$$\cos 2A = \cos^2 A - \sin^2 A$$
$$= (1 - \sin^2 A) - \sin^2 A$$
$$= 1 - 2 \sin^2 A$$

so $\qquad \cos 2A = 1 - 2 \sin^2 A. \qquad \text{Equation 11}$

You can also write Equations 10 and 11 in the forms

$$1 + \cos 2A = 2 \cos^2 A \qquad \text{Equation 12}$$

and $\qquad 1 - \cos 2A = 2 \sin^2 A. \qquad \text{Equation 13}$

If you write Equations 9, 10 and 11 in half-angle form, you get

$$\cos \theta = \cos^2 \tfrac{1}{2}\theta - \sin^2 \tfrac{1}{2}\theta, \qquad \text{Equation 14}$$
$$\cos \theta = 2 \cos^2 \tfrac{1}{2}\theta - 1, \qquad \text{Equation 15}$$

and $\qquad \cos \theta = 1 - 2 \sin^2 \tfrac{1}{2}\theta. \qquad \text{Equation 16}$

If you put $B = A$ in the formula

$$\tan(A + B) = \frac{\tan A + \tan B}{1 - \tan A \tan B},$$

you obtain

$$\tan 2A = \frac{\tan A + \tan A}{1 - \tan A \tan A}$$

$$= \frac{2 \tan A}{1 - \tan^2 A}$$

so

$$\tan 2A = \frac{2 \tan A}{1 - \tan^2 A}. \qquad \text{❋Equation 17}$$

In half-angle form, this is

$$\tan \theta = \frac{2 \tan \frac{1}{2}\theta}{1 - \tan^2 \frac{1}{2}\theta}. \qquad \text{Equation 18}$$

Exercise 10.2

1 Given that $\sin A = \frac{3}{5}$, and that A is acute, find the values of $\sin 2A$, $\cos 2A$ and $\tan 2A$.

2 Given that $\sin A = \frac{3}{5}$, and that A is obtuse, find the values of $\sin 2A$, $\cos 2A$ and $\tan 2A$.

3 Find $\sin 2\theta$, $\cos 2\theta$ and $\tan 2\theta$ when $\sin \theta = 0.25$ and θ is acute.

4 Given the values of $\sin 45°$ and $\cos 45°$, use the formulae of the previous sections to calculate $\sin 90°$ and $\cos 90°$.

5 Given that $\cos B = 0.66$, and that B is acute, find the values of $\sin 2B$ and $\cos 2B$.

6 Given that $\cos B = 0.66$, and that B is not acute, find the values of $\sin 2B$ and $\cos 2B$.

7 Find the values of $2\sin 36° \cos 36°$ and $2 \cos^2 36° - 1$.

8 Given that $\cos 2A = \frac{3}{5}$, find the two possible values of $\tan A$.

9 Prove that $\sin \frac{1}{2}\theta = \pm\sqrt{\dfrac{1 - \cos \theta}{2}}$ and $\cos \frac{1}{2}\theta = \pm\sqrt{\dfrac{1 + \cos \theta}{2}}$.

10 Given that $\cos \theta = \frac{1}{2}$, find $\sin \frac{1}{2}\theta$ and $\cos \frac{1}{2}\theta$.

11 Given that $\cos 2\theta = 0.28$, find $\sin \theta$.

12 Find the value of $\sqrt{\dfrac{1 - \cos 40°}{1 + \cos 40°}}$.

10.7 Identities

It is often extremely useful to be able to simplify a trigonometric expression, or to be able to prove that two expressions are equal for all possible values of the angle or angles involved.

An equation which is true for all possible values of the angle or angles is called an **identity**.

For example, $\sin(A - B) = \sin A \cos B - \cos A \sin B$ is an example of an identity, as are all the formulae given in Equations 1 to 18. So also is $1 + \sin\theta = (\sin\frac{1}{2}\theta + \cos\frac{1}{2}\theta)^2$, but the latter needs to be proved to be an identity.

To prove that a trigonometric equation is an identity, you can choose one of two possible methods.

Method 1 Start with the side of the identity you believe to be the more complicated, and manipulate it, using various formulae including those in Equations 1 to 18, until you arrive at the other side.

Method 2 If you do not see how to proceed with Method 1, then it may help to take the right-hand side from the left-hand side and to prove that the result is zero.

Here are some examples.

Example 10.4

Prove the identity $1 + \sin\theta = (\sin\frac{1}{2}\theta + \cos\frac{1}{2}\theta)^2$.

The more complicated side is the right-hand side, so the strategy will be to use Method 1 and to prove that this is equal to the left-hand side.

In the work which follows, LHS will be used to denote the left-hand side of an equation and RHS the right-hand side.

$$\text{RHS} = \sin^2\tfrac{1}{2}\theta + 2\sin\tfrac{1}{2}\theta\cos\tfrac{1}{2}\theta + \cos^2\tfrac{1}{2}\theta$$
$$= (\sin^2\tfrac{1}{2}\theta + \cos^2\tfrac{1}{2}\theta) + 2\sin\tfrac{1}{2}\theta\cos\tfrac{1}{2}\theta.$$

Now use Pythagoras's theorem, and Equation 8. Then

$$\text{RHS} = 1 + 2\sin\tfrac{1}{2}\theta\cos\tfrac{1}{2}\theta$$
$$= 1 + \sin\theta = \text{LHS}.$$

Since RHS = LHS, the identity is true. ∎

Example 10.5

Prove the identity $\dfrac{\sin A}{1 - \cos A} = \dfrac{1 + \cos A}{\sin A}$.

It is not clear which side is the more complicated, so use Method 2. The advantage with Method 2 is that there is then an obvious way to proceed, that is, change the resulting expression for LHS − RHS into a single fraction.

$$\begin{aligned}
\text{LHS} - \text{RHS} &= \frac{\sin A}{1 - \cos A} - \frac{1 + \cos A}{\sin A} \\
&= \frac{\sin^2 A - (1 - \cos A)(1 + \cos A)}{\sin A(1 - \cos A)} \\
&= \frac{\sin^2 A - (1 - \cos^2 A)}{\sin A(1 - \cos A)} \\
&= \frac{\sin^2 A - 1 + \cos^2 A}{\sin A(1 - \cos A)} \\
&= 0.
\end{aligned}$$

The last step follows from Pythagoras's theorem,

$$\sin^2 A + \cos^2 A = 1.$$

Since RHS = LHS, the identity is true. ∎

Example 10.6

Prove the identity $\cos^4 \phi - \sin^4 \phi = \cos 2\phi$.

Starting from the left-hand side, and using Method 1,

$$\begin{aligned}
\text{LHS} &= \cos^4 \phi - \sin^4 \phi \\
&= (\cos^2 \phi - \sin^2 \phi)(\cos^2 \phi + \sin^2 \phi) \\
&= \cos 2\phi \times 1 \\
&= \cos 2\phi = \text{RHS}.
\end{aligned}$$

Equation 9 and Pythagoras's theorem are used in the second step of the argument.

Since RHS = LHS, the identity is true. ∎

You must be careful not to use an illogical argument when proving identities. Here is an example of an illogical argument.

Example 10.7 An illogical 'proof'

Prove that 2 = 3.

If $2 = 3$

then $3 = 2$.

Adding the left-hand sides and right-hand sides of these equations gives

$$5 = 5.$$

As this is true, the original statement is true, so $2 = 3$. ∎

This is obviously absurd, so the argument itself must be invalid. Nevertheless, many students use just this argument when attempting to prove something which is true. Beware! However, if you stick to the arguments involved with Methods 1 and 2, you will be safe.

Exercise 10.3

In questions 1 to 7, prove the given identities.

1 $\sin(A + B) + \sin(A - B) = 2\sin A \cos B$

2 $\cos(A + B) + \cos(A - B) = 2\cos A \cos B$

3 $\dfrac{\cos 2A}{\cos A + \sin A} = \cos A - \sin A$

4 $\sin 3A = 3\sin A - 4\sin^3 A$

5 $\dfrac{\sin A}{\cos A} + \dfrac{\cos A}{\sin A} = \dfrac{2}{\sin 2A}$

6 $\dfrac{1}{\sin\theta} + \dfrac{\cos\theta}{\sin\theta} = \dfrac{\cos\frac{1}{2}\theta}{\sin\frac{1}{2}\theta}$

7 $\dfrac{\cos\theta}{\sin\phi} + \dfrac{\cos\theta}{\cos\phi} = \dfrac{2\sin(\theta + \phi)}{\sin 2\phi}$

10.8 More trigonometric equations

Sometimes you can use some of the formulae on earlier pages to help you to solve equations. Here are some examples.

Example 10.8

Solve the equation $\cos 2\theta° = \sin\theta°$ giving all solutions between −180 and 180 inclusive.

You can replace the $\cos 2\theta°$ term by $1 - 2\sin^2\theta°$ (Equation 11) and you will then have an equation in $\sin\theta$.

Then $$\cos 2\theta° = \sin \theta°$$

so $$1 - 2\sin^2\theta° = \sin\theta°$$

so $$2\sin^2\theta° + \sin\theta° - 1 = 0.$$

Factorizing $$(2\sin\theta° - 1)(\sin\theta° + 1) = 0$$

so $$\sin\theta° = 0.5 \text{ or } \sin\theta° = -1.$$

Using the methods of Section 6.2, you can solve these equations for θ.

When $\sin\theta° = 0.5$, $\theta = 30$ or 150, and when $\sin\theta° = -1$, $\theta = -90$.

Therefore $\theta = -90$, 30 or 150. ∎

Exercise 10.4

In questions 1 to 10, solve the given equation for θ, giving your answers in the interval from -180 to 180.

1 $\sin 2\theta° = \cos\theta°$

2 $2\cos^2 x° - 1 = \frac{1}{2}$

3 $\cos 2\theta° = \sin\theta° \cos\theta°$

4 $4\sin\theta° \cos\theta° = 1$

5 $1 - 2\sin^2\theta° = 2\sin\theta° \cos\theta°$

6 $\dfrac{1-\tan^2\theta°}{1+\tan^2\theta°} = \frac{1}{2}$

7 $\cos 2\theta° = \cos\theta°$

8 $3\sin\theta° = 4\sin^3\theta°$

9 $4\cos^3\theta° = 3\cos\theta°$

10 $2\tan\theta° = 1 - \tan^2\theta°$

11
the form
a sin x + b cos x

In this chapter you will learn:
- that the graph of
  $y = a \sin x + b \cos x$ is like the
  graph of a sine or a cosine
- how to express
  $a \sin x + b \cos x$ in the form
  $R \sin(x + \alpha)$, and find R and α
  in terms of a and b
- how to use the form
  $R \sin(x + \alpha)$ in applications.

11.1 Introduction

If you have a graphics calculator available, try drawing the graphs of functions of the form $y = 2 \sin x + 3 \cos x$ and $y = 3 \sin x - 4 \cos x$. These two graphs are shown in Figure 11.1.

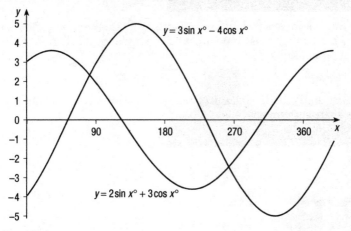

figure 11.1

Both graphs have the characteristic wave properties of the sine and cosine functions. They have been enlarged in the y-direction, by different amounts, and translated in the x-direction, by different amounts.

This suggests that you may be able to write both of these functions in the form

$$y = R \sin(x + \alpha)°$$

for suitable values of the constants R and α, where the value of R is positive.

This idea is pursued in the next section.

11.2 The form $y = a \sin x + b \cos x$

If you try to choose the values of R and α so that the function $y = R \sin(x + \alpha)$ is identical with $y = a \sin x + b \cos x$, you can start by expanding $\sin(x + \alpha)$ so that

$$y = R \sin x \cos \alpha + R \cos x \sin \alpha$$

or $\qquad y = (R \cos \alpha) \sin x + (R \sin \alpha) \cos x.$

If this is the same function as $y = a \sin x + b \cos x$ for all values of x then

$$R \cos \alpha = a \text{ and } R \sin \alpha = b,$$

that is $$\cos \alpha = \frac{a}{R} \text{ and } \sin \alpha = \frac{b}{R}.$$

You can interpret these two equations by thinking of a and b as the adjacent and opposite of a triangle which has R as its hypotenuse.

Therefore $$R = \sqrt{a^2 + b^2}.$$

Alternatively, if you square these equations and add the two equations, you get

$$R^2 \cos^2 \alpha + R^2 \sin^2 \alpha = a^2 + b^2,$$

so $$R^2(\cos^2 \alpha + \sin^2 \alpha) = a^2 + b^2,$$

that is $$R^2 = a^2 + b^2 \text{ and } R = \pm \sqrt{a^2 + b^2}.$$

The value of R is always chosen to be positive.

Therefore $$R = \sqrt{a^2 + b^2}.$$

Then $$\cos \alpha = \frac{a}{\sqrt{a^2 + b^2}} \text{ and } \sin \alpha = \frac{b}{\sqrt{a^2 + b^2}}.$$

These three equations,

$$R = \sqrt{a^2 + b^2} \qquad \textbf{Equation 1}$$

and $$\cos \alpha = \frac{a}{\sqrt{a^2 + b^2}} \text{ and } \sin \alpha = \frac{b}{\sqrt{a^2 + b^2}} \quad \textbf{Equations 2}$$

enable you to determine the values of R and α.

Example 11.1

Express $2 \sin x° + 3 \cos x°$ in the form $R \sin(x + \alpha)°$, where the angles are in degrees.

For the function $2 \sin x° + 3 \cos x°$, $a = 2$ and $b = 3$. From Equation 1, $R = \sqrt{13}$. Using this value in Equation 2,

$$\cos \alpha° = \frac{2}{\sqrt{13}} \text{ and } \sin \alpha° = \frac{3}{\sqrt{13}}.$$

These equations, in which $\cos\alpha$ and $\sin\alpha$ are both positive, show that α is a first-quadrant angle, and that

$$\alpha \approx 56.31.$$

Therefore $\qquad 2\sin x° + 3\cos x° \equiv \sqrt{13}\sin(x + \alpha)$

where $\qquad\qquad \alpha \approx 56.31.$ ∎

Note that the symbol '≡' is used to mean 'identically equal to'.

Example 11.2

Express $3\sin x - 4\cos x$ in the form $R\sin(x + \alpha)$ with α in radians.

For the function $3\sin x - 4\cos x$, $a = 3$ and $b = -4$. From Equation 1, $R = 5$. Using this value in Equations 2,

$$\cos\alpha = \frac{3}{5} \text{ and } \sin\alpha = \frac{-4}{5}.$$

These equations, in which $\cos\alpha$ is positive and $\sin\alpha$ is negative, show that α is a fourth-quadrant angle, and that:

$$\alpha \approx 5.36.$$

Therefore $\qquad 3\sin x - 4\cos x \equiv 5\sin(x + \alpha)$

where $\qquad\qquad \alpha \approx 5.36.$ ∎

Example 11.3

Express $\sin x - 2\cos x$ in the form $R\cos(x + \alpha)$, where $R > 0$.

This is a different form from the original, but it is not difficult to adapt the methods of the beginning of this section.

Comparing the expanded form $R\cos x\cos\alpha - R\sin x\sin\alpha$ with the form $\sin x - 2\cos x$, gives

$$R\cos\alpha = 2 \text{ and } R\sin\alpha = 1.$$

Squaring and adding, as before, gives

$$R = \sqrt{5}$$

and $\qquad\qquad \cos\alpha = \frac{2}{\sqrt{5}} \text{ and } \sin\alpha = \frac{1}{\sqrt{5}}.$

Thus, as $\cos\alpha$ and $\sin\alpha$ are both positive, α is a first-quadrant angle, and

$$\alpha \approx 0.46.$$

Thus
$$\sin x - 2\cos x \equiv \sqrt{5}\,\cos(x + \alpha)$$

where
$$\alpha \approx 0.46. \ \blacksquare$$

Exercise 11.1
In questions 1 to 6, write the given function in the form $R\sin(x + \alpha)°$.

1 $\sin x° + \cos x°$ **2** $5\sin x° + 12\cos x°$

3 $2\cos x° + 5\sin x°$ **4** $\cos x° - \sin x°$

5 $\sin x° - 3\cos x°$ **6** $3\cos x° - \sin x°$

In questions 7 to 9, give α in radians.

7   Write the function $\sin x + \cos x$ in the form $R\cos(x + \alpha)$.

8   Write the function $\sin x + \cos x$ in the form $R\cos(x - \alpha)$.

9   Write the function $\sin x + \cos x$ in the form $R\sin(x - \alpha)$.

10   By writing the functions $7\cos x + \sin x$ and $5\cos x - 5\sin x$ in the form $R\sin(x + \alpha)$, show that they have the same maximum value.

11.3 Using the alternative form

There are two main advantages in writing something like $\sin x + \cos x$ in any one of the four forms $R\sin(x + \alpha)$, $R\sin(x - \alpha)$, $R\cos(x + \alpha)$ and $R\cos(x - \alpha)$.

It enables you to solve equations easily, and to find the maximum and minimum values of the function without further work.

Here are some examples.

Example 11.4
Solve the equation $5\sin x° + 8\cos x° = 3$ giving all the solutions between 0 and 360.

Using the method of Section 11.2, you can write $5\sin x° + 8\cos x°$ as

$$\sqrt{89}\,\sin(x + 57.99)°.$$

The equation then becomes

$$\sqrt{89}\,\sin(x + 57.99)° = 3$$

or $$\sin(x + 57.99)° = \frac{3}{\sqrt{89}}.$$

Let $z = x + 57.99$. Then you require the solutions of $\sin z° = \frac{3}{\sqrt{89}}$ for values of z between 57.99 and 417.99.

The principal angle is 18.54, and the other angle between 0 and 360 is the second quadrant angle (see Section 6.2), $180 - 18.54 = 161.46$.

You now have to add 360 to the first of these to find the value of z in the required interval. Then the two solutions for z are

$$z = 378.54 \text{ and } 161.46.$$

You find the solutions for x by substituting $z = x + 57.99$, and you find that

$$x = 378.54 - 57.99 = 320.55$$

or $$x = 161.46 - 57.99 = 103.47.$$

Thus the solutions are 103.47 and 320.55°. ∎

Example 11.5

Find the maximum and minimum values of $\sin x - 3 \cos x$ and the values of x, in radians, for which they occur.

Writing $\sin x - 3 \cos x$ in the form $R \sin(x - \alpha)$ using the methods in Section 11.2 gives

$$\sin x - 3 \cos x \equiv \sqrt{10} \sin(x - 1.249).$$

The question now becomes: Find the maximum and minimum values of $\sqrt{10} \sin(x - 1.249)$ and the values of x for which they occur.

You know that the maximum of a sine function is 1 and that it occurs when the angle is $\frac{1}{2}\pi$.

Thus the maximum value of $\sqrt{10} \sin(x - 1.249)$ is $\sqrt{10}$, and this occurs when $x - 1.249 = \frac{1}{2}\pi$, that is, when $x = 2.820$.

Similarly the minimum value of a sine function is -1, and this occurs when the angle is $\frac{2}{3}\pi$.

Thus the minimum value of $\sqrt{10} \sin(x - 1.259)$ is $-\sqrt{10}$, and this occurs when $x - 1.259 = \frac{3}{2}\pi$, that is, when $x = 5.961$. ∎

Example 11.6

Solve the equation $3 \cos 2x° - 4 \sin 2x° = 2$ giving all solutions in the interval −180 to 180.

Begin by writing $y = 2x$: the equation becomes

$$3 \cos y° - 4 \sin y° = 2$$

with solutions for y needed in the interval −360 to 360.

Writing $3 \cos y° - 4 \sin y°$ in the form $R \sin(y - \alpha)°$ using the methods in Section 11.2 gives

$$3 \cos y° - 4 \sin y° \equiv 5 \cos(y + 53.13)°.$$

Solving the equation $5 \cos(y + 53.13)° = 2$ gives

$$\cos(y + 53.13)° = 0.4.$$

The principal angle is 66.42.

Using the methods of Section 6.3, the angles between −360 and 360 satisfying this equation are

$$-293.58, -66.42, 66.42, 293.58.$$

Thus $y + 53.13 = -293.58, -66.42, 66.42, 293.58$

so $y = -346.71, -119.55, 13.29, 240.45.$

Finally, dividing by 2 as $y = 2x$ gives

$$x = -173.35, -59.78, 6.65, 120.22.$$

Note that the decision about whether to round up or to round down the final figure on dividing by 2 was made by keeping more significant figures on the calculator. ∎

Example 11.7

Show that the equation $2 \sin x + 3 \cos x = 4$ has no solutions.

You can write this equation in the form

$$\sqrt{13} \sin(x + \alpha) = 4$$

for a suitable value of α.

You can then rewrite the equation in the form

$$\sin(x + \alpha) = \frac{4}{\sqrt{13}}.$$

As $\frac{4}{\sqrt{13}} > 1$, there is no solution to this equation. ∎

Exercise 11.2

In questions 1 to 6, solve the given equation for θ, giving the value of θ in the interval 0 to 360 inclusive.

1 $\sin\theta° + \cos\theta° = 1$

2 $\sin\theta° + \sqrt{3}\cos\theta° = 1$

3 $3\cos\theta° - 2\sin\theta° = 1$

4 $12\sin\theta° - 5\cos\theta° = 5$

5 $-8\cos\theta° - 7\sin\theta° = 5$

6 $\cos2\theta° - \sin2\theta° = -1$

In questions 7 to 12, solve the given equation for θ, giving the value of θ in the interval -180 to 180 inclusive.

7 $\cos\theta° + \sin\theta° = -1$

8 $\sqrt{3}\sin\theta° + \cos\theta° = -1$

9 $3\cos\theta° - \sin\theta° = 2$

10 $-2\cos\theta° - 3\sin\theta° = 3$

11 $6\sin\theta° - 7\cos\theta° = -8$

12 $\sqrt{3}\cos2\theta° - \sin2\theta° = -1$

In questions 13 to 18, find the maximum and minimum values of the function and the values of x in the interval between -180 and 180 inclusive, for which they occur.

13 $y = 2\sin x° - \cos x°$

14 $y = 3\cos x° - 4\sin x°$

15 $y = \sqrt{3}\cos2x° - \sin2x°$

16 $y = \cos2x° - \sin2x°$

17 $y = 3\sin x° + 4\cos x° + 2$

18 $y = \sqrt{2}\cos2x° - \sin2x° + 3$

12

the factor formulae

In this chapter you will learn:

- how to express the sum and difference of two sines or cosines in an alternative form as a product
- how to do this process in reverse
- how to use both the processes in solving problems.

12.1 The first set of factor formulae

In Exercise 10.3, question 1, you were asked to prove the identity

$$\sin(A + B) + \sin(A - B) = 2 \sin A \cos B.$$

The proof of this identity relies on starting with the left-hand side and expanding the terms using Equations 1 and 2 of Chapter 10 to get

$$\sin(A + B) = \sin A \cos B + \cos A \sin B$$

and $\qquad \sin(A - B) = \sin A \cos B - \cos A \sin B.$

Adding the left-hand sides of these two equations you obtain the required result

$$\sin(A + B) + \sin(A - B) = 2 \sin A \cos B,$$

which will be used in the rewritten form

$$2 \sin A \cos B = \sin(A + B) + \sin(A - B). \quad \textbf{Equation 1}$$

This is the first formula of its type. These formulae enable you to move from a product of sines and cosines, to a sum or difference of sines and cosines equal to it.

Example 12.1
Use Equation 1 to simplify $2 \sin 30° \cos 60°$.

$$\begin{aligned}
2 \sin 30° \cos 60° &= \sin(30 + 60)° + \sin(30 - 60)° \\
&= \sin 90° + \sin(-30)° \\
&= 1 - \sin 30° \\
&= 1 - \tfrac{1}{2} = \tfrac{1}{2}. \ \blacksquare
\end{aligned}$$

In the middle of the example, the fact that $\sin(-\theta) = - \sin \theta$ for any angle θ was used to change $\sin(-30)°$ to $\sin 30°$.

If you subtract the equations

$$\sin(A + B) = \sin A \cos B + \cos A \sin B$$

and $\qquad \sin(A - B) = \sin A \cos B - \cos A \sin B$

you obtain $\quad \sin(A + B) - \sin(A - B) = 2 \cos A \sin B,$

that is $\qquad 2 \cos A \sin B = \sin(A + B) - \sin(A - B). \quad \textbf{Equation 2}$

If you had used Equation 2 to solve Example 12.1, you would say

$$2 \sin 30° \cos 60° = 2 \cos 60° \sin 30°$$
$$= \sin(60 + 30)° - \sin(60 - 30)°$$
$$= \sin 90° - \sin 30°$$
$$= 1 - \sin 30°$$
$$= 1 - \tfrac{1}{2} = \tfrac{1}{2}.$$

Two other formulae come from the equivalent formulae for $\cos(A + B)$ and $\cos(A - B)$,

$$\cos(A + B) = \cos A \cos B - \sin A \sin B$$

and $\qquad \cos(A - B) = \cos A \cos B + \sin A \sin B.$

First adding, and then subtracting, these equations gives

$$\cos(A + B) + \cos(A - B) = 2 \cos A \cos B$$

and $\qquad \cos(A + B) - \cos(A - B) = -2 \sin A \sin B.$

When you rewrite these equations you have

$$2 \cos A \cos B = \cos(A + B) + \cos(A - B) \quad \textbf{Equation 3}$$

and $\qquad 2 \sin A \sin B = \cos(A - B) - \cos(A + B). \quad \textbf{Equation 4}$

Note the form of these four equations which are gathered together for convenience

$$2 \sin A \cos B = \sin(A + B) + \sin(A - B) \quad \textbf{Equation 1}$$
$$2 \cos A \sin B = \sin(A + B) - \sin(A - B) \quad \textbf{Equation 2}$$
$$2 \cos A \cos B = \cos(A + B) + \cos(A - B) \quad \textbf{Equation 3}$$
$$2 \sin A \sin B = \cos(A - B) - \cos(A + B). \quad \textbf{Equation 4}$$

These formulae are often remembered as

$$2 \times \sin \times \cos = \sin(\textit{sum}) + \sin(\textit{difference}),$$
$$2 \times \cos \times \sin = \sin(\textit{sum}) - \sin(\textit{difference}),$$
$$2 \times \cos \times \cos = \cos(\textit{sum}) + \cos(\textit{difference}),$$
$$2 \times \sin \times \sin = \cos(\textit{difference}) - \cos(\textit{sum}).$$

Note the following points.

- In Equations 1 and 2, it is important that the 'difference' is found by subtracting B from A.

- In Equations 3 and 4 it is not important whether you take the difference as $A - B$ or as $B - A$; the equation $\cos(-\theta) = \cos\theta$ for all angles θ ensures that $\cos(B - A) = \cos(A - B)$.

- The order of the right-hand side in Equation 4 is different from the other formulae.

Example 12.2

Express $\sin 5\theta \cos 3\theta$ as the sum of two trigonometric ratios.

Using Equation 1, $2 \sin A \cos B = \sin(A + B) + \sin(A + B)$, gives

$$\sin 5\theta \cos 3\theta = \tfrac{1}{2}(2 \sin 5\theta \cos 3\theta)$$
$$= \tfrac{1}{2}(\sin(5\theta + 3\theta) + \sin(5\theta - 3\theta))$$
$$= \tfrac{1}{2}(\sin(8\theta) + \sin(2\theta)). \ \blacksquare$$

Example 12.3

Change $\sin 70° \sin 20°$ into a sum.

Using Equation 4, $2 \sin A \sin B = \cos(A - B) - \cos(A + B)$, gives

$$\sin 70° \cos 20° = \tfrac{1}{2}(2 \sin 70° \cos 20°)$$
$$= \tfrac{1}{2}(\cos(70° - 20°) - \cos(70° + 20°))$$
$$= \tfrac{1}{2}(\cos 50° - \cos 90°)$$
$$= \tfrac{1}{2}(\cos 50° - 0)$$
$$= \tfrac{1}{2}\cos 50°. \ \blacksquare$$

Exercise 12.1

In questions 1 to 8, express the given expression as the sum or difference of two trigonometric ratios.

1   $\sin 3\theta \cos\theta$ 2 $\sin 35° \cos 45°$

3   $\cos 50° \cos 30°$ 4 $\cos 5\theta \sin 3\theta$

5   $\cos(C + 2D) \cos(2C + D)$ 6 $\cos 60° \sin 30°$

7   $2 \sin 3A \sin A$ 8 $\cos(3C + 5D) \sin(3C - 5D)$

9   In Equation 1, put $A = 90 - C$ and $B = 90 - D$ and simplify both sides of the resulting identity. What equation results?

10   In Equation 1, put $A = 90 - C$ and simplify both sides of the resulting identity. What equation results?

12.2 The second set of factor formulae

The second set of factor formulae are really a rehash of the first set which enable you to work the other way round, that is, to write the sum or difference of two sines or two cosines as a product of sines and cosines.

Starting from

$$2 \sin A \cos B = \sin(A + B) + \sin(A - B),$$

write it the other way round as

$$\sin(A + B) + \sin(A - B) = 2 \sin A \cos B.$$

Put $A + B = C$ and $A - B = D$.

Then the identity becomes

$$\sin C + \sin D = 2 \sin A \cos B.$$

If you can write A and B in terms of C and D you will obtain a formula for the sum of two sines.

From the equations

$$A + B = C$$

and $$A - B = D,$$

you can use simultaneous equations to deduce that

$$A = \frac{C + D}{2} \text{ and } B = \frac{C - D}{2}.$$

Then $$\sin C + \sin D = 2 \sin \frac{C + D}{2} \cos \frac{C - D}{2}. \quad \textbf{Equation 5}$$

You can deduce a second formula in a similar way from Equation 2, by a similar method.

Equation 2 then becomes

$$\sin C - \sin D = 2 \sin \frac{C - D}{2} \cos \frac{C + D}{2}. \quad \textbf{Equation 6}$$

Similar methods applied to Equations 3 and 4 give

$$\cos C + \cos D = 2 \cos \frac{C + D}{2} \cos \frac{C - D}{2} \quad \textbf{Equation 7}$$

and $$\cos C - \cos D = 2 \sin \frac{C + D}{2} \sin \frac{D - C}{2}. \quad \textbf{Equation 8}$$

Equations 5 to 8 are often remembered as

$\sin + \sin = 2 \times \sin(semisum) \times \cos(semidifference)$,

$\sin - \sin = 2 \times \cos(semisum) \times \sin(semidifference)$,

$\cos + \cos = 2 \times \cos(semisum) \times \cos(semidifference)$,

$\cos - \cos = 2 \times \sin(semisum) \times \sin(semidifference\ reversed)$.

Example 12.4
Transform $\sin 25° + \sin 18°$ into a product.

Using Equation 5, $\sin C + \sin D = 2 \sin \dfrac{C+D}{2} \cos \dfrac{C-D}{2}$, gives

$$\sin 25° + \sin 18° = 2 \sin \frac{25° + 18°}{2} \cos \frac{25° - 18°}{2}$$

$$= 2 \sin 21.5° \cos 3.5°. \blacksquare$$

Example 12.5
Change $\cos 3\theta - \cos 7\theta$ into a product.

Using Equation 8, $\cos C - \cos D = 2 \sin \dfrac{C+D}{2} \sin \dfrac{D-C}{2}$, gives

$$\cos 3\theta - \cos 7\theta = 2 \sin \frac{3\theta + 7\theta}{2} \sin \frac{7\theta - 3\theta}{2}$$

$$= 2 \sin 5\theta \sin 2\theta. \blacksquare$$

Example 12.6
Solve the equation $\sin \theta° - \sin 3\theta°$, giving solutions in the interval -180 to 180.

$$\sin \theta° - \sin 3\theta° = 2 \cos 2\theta° \sin(-\theta)°$$

$$= -2 \cos 2\theta° \sin \theta°.$$

Hence $\qquad \cos 2\theta° = 0$ or $\sin \theta° = 0$.

The solutions of $\cos 2\theta° = 0$ are $-135, -45, 45, 135$ and the solutions of $\sin \theta° = 0$ are $-180, 0, 180$.

Therefore the solutions of the original equation are

$$-180, -135, -45, 0, 45, 135, 180. \blacksquare$$

You could also have expanded $\sin 3\theta$ in the form $3\sin\theta - 4\sin^3\theta$ and then solved the equation $4\sin^3\theta - 2\sin\theta = 0$ to get the same result.

Example 12.7
Prove that if $A + B + C = 180$, then

$$\frac{\sin A + \sin B - \sin C}{\sin A + \sin B + \sin C} = \tan A \tan B.$$

$$\text{LHS} = \frac{(\sin A + \sin B) - \sin C}{(\sin A + \sin B) + \sin C}$$

$$= \frac{2\sin\frac{1}{2}(A+B)\cos\frac{1}{2}(A-B) - 2\sin\frac{1}{2}C\cos\frac{1}{2}C}{2\sin\frac{1}{2}(A+B)\cos\frac{1}{2}(A-B) + 2\sin\frac{1}{2}C\cos\frac{1}{2}C}$$

$$= \frac{2\cos\frac{1}{2}C\cos\frac{1}{2}(A-B) - 2\sin\frac{1}{2}C\cos\frac{1}{2}C}{2\cos\frac{1}{2}C\cos\frac{1}{2}(A-B) + 2\sin\frac{1}{2}C\cos\frac{1}{2}C}$$

$$= \frac{2\cos\frac{1}{2}C(\cos\frac{1}{2}(A-B) - \sin\frac{1}{2}C)}{2\cos\frac{1}{2}C(\cos\frac{1}{2}(A-B) + \sin\frac{1}{2}C)}$$

$$= \frac{\cos\frac{1}{2}(A-B) - \sin\frac{1}{2}C}{\cos\frac{1}{2}(A-B) + \sin\frac{1}{2}C}$$

$$= \frac{\cos\frac{1}{2}(A-B) - \cos\frac{1}{2}(A+B)}{\cos\frac{1}{2}(A-B) + \cos\frac{1}{2}(A+B)}$$

$$= \frac{2\sin A \sin B}{2\cos A \cos B} = \tan A \tan B = \text{RHS}.$$

Since the LHS = RHS, the identity is true. ∎

Notice how in Example 12.7, the facts that $\cos(90 - \theta)° = \sin\theta°$
and $\sin(90 - \theta)° = \cos\theta°$ were used in the forms

$$\cos\tfrac{1}{2}(A + B) = \sin\tfrac{1}{2}C \text{ and } \sin\tfrac{1}{2}(A + B) = \cos\tfrac{1}{2}C.$$

Exercise 12.2

In questions 1 to 6, express the given sum or difference as the product of two trigonometric ratios.

1 $\sin 4A + \sin 2A$ **2** $\sin 5A - \sin A$

3 $\cos 4\theta - \cos 2\theta$ **4** $\cos A - \cos 5A$

5 $\cos 47° + \cos 35°$ **6** $\sin 49° - \sin 23°$

In questions 7 to 10, use the factor formulae to simplify the given expressions.

7 $\dfrac{\sin 30° + \sin 60°}{\cos 30° - \cos 60°}$ **8** $\dfrac{\sin \alpha + \sin \beta}{\cos \alpha + \cos \beta}$

9 $\dfrac{\cos \theta - \cos 3\theta}{\sin \theta + \sin 3\theta}$ **10** $\dfrac{\sin \theta + 2\sin 2\theta + \sin 3\theta}{\cos \theta + 2\cos 2\theta + \cos 3\theta}$

In questions 11 to 14, use the factor formulae to solve the following equations, giving all solutions in the interval 0 to 360.

11 $\cos \theta° - \cos 2\theta° = 0$ **12** $\sin \theta° + \sin 2\theta° = 0$

13 $\sin \theta° + \sin 2\theta° + \sin 3\theta° = 0$ **14** $\cos \theta° + 2\cos 2\theta° + \cos 3\theta° = 0$

In questions 15 to 17, you are given that $A + B + C = 180°$. Prove that each of the following results is true.

15 $\sin A + \sin B + \sin C = 4\cos \tfrac{1}{2}A \cos \tfrac{1}{2}B \cos \tfrac{1}{2}C$

16 $\sin A + \sin B - \sin C = 4\sin \tfrac{1}{2}A \sin \tfrac{1}{2}B \cos \tfrac{1}{2}C$

17 $\cos A + \cos B + \cos C - 1 = 4\sin \tfrac{1}{2}A \sin \tfrac{1}{2}B \sin \tfrac{1}{2}C$

13

circles related to a triangle

In this chapter you will learn:
- that the circumcircle is the circle which passes through the vertices of a triangle
- that the incircle and the three ecircles all touch the three sides of a triangle
- how to calculate the radii of these circles from information about the triangle.

13.1 The circumcircle

The circumcircle of a triangle *ABC* (see Figure 13.1) is the circle which passes through each of the vertices. The centre of the circumcircle will be denoted by *O* and its radius will be denoted by *R*.

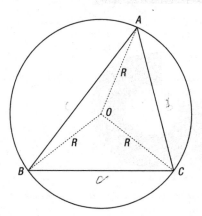

figure 13.1

To calculate the radius *R* of the circumcircle, drop the perpendicular from *O* on to the side *BC* to meet *BC* at *N*. See Figure 13.2.

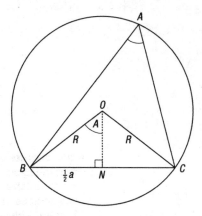

figure 13.2

As *BOC* is an isosceles triangle, because two of its sides are *R*, the line *ON* bisects the base. Therefore $BN = NC = \frac{1}{2}a$.

In addition, angle BOC, the angle at the centre of the circle standing on the arc BC, is twice the angle BAC, which stands on the same arc. Thus angle $BOC = 2A$, so angle $BON = A$.

Therefore, in triangle BON,

$$\sin A = \frac{\frac{1}{2}a}{R}$$

so

$$2R = \frac{a}{\sin A}.$$

You will recognize that the right-hand side of this formula for R also occurs in the sine formula.

Therefore

$$2R = \frac{a}{\sin A} = \frac{b}{\sin B} = \frac{c}{\sin C}. \qquad \textbf{Equation 1}$$

Equation 1, which enables you to calculate the radius R of the circumcircle, is what some people understand by the sine formula for a triangle, rather than the shorter version on page 66.

Note that the equation $2R = \dfrac{a}{\sin A}$ would itself give you a proof of the sine formula for the triangle. For, by symmetry, the radius of the circumcircle must be a property of the triangle as a whole, and not 'biased' towards one particular vertex. Therefore, if you had started with another vertex, you would obtain the formulae $2R = \dfrac{b}{\sin B}$ or $2R = \dfrac{c}{\sin C}$.

Thus $2R = \dfrac{a}{\sin A} = \dfrac{b}{\sin B} = \dfrac{c}{\sin C}$.

Example 13.1

Find the radius of the circumcircle of the triangle with sides of length 4 cm, 5 cm and 6 cm.

Let $a = 4$, $b = 5$ and $c = 6$. Then calculate an angle of the triangle, say the largest angle, C, using the cosine formula. From the value of $\cos C$, work out the value of $\sin C$, and then use Equation 1 to find R.

Using the cosine formula, $c^2 = a^2 + b^2 - 2ab \cos C$,

$$6^2 = 4^2 + 5^2 - 2 \times 4 \times 5 \times \cos C$$

which reduces to $\qquad 40 \cos C = 5$

or $\qquad\qquad\qquad \cos C = \frac{1}{8}.$

Using $\sin^2 C = 1 - \cos^2 C$,

$$\sin^2 C = 1 - \frac{1}{64} = \frac{63}{64}$$

so
$$\sin C = \sqrt{\frac{63}{64}} = \frac{3\sqrt{7}}{8}.$$

Finally, using the full version of the sine formula, Equation 1, $2R = \dfrac{c}{\sin C}$, you obtain

$$2R = \frac{6}{\frac{3\sqrt{7}}{8}} = \frac{16}{\sqrt{7}}.$$

Thus the radius R of the circumcircle is $\dfrac{8}{\sqrt{7}}$ cm. ∎

Example 13.2
The angles A, B and C of a triangle are 50°, 60° and 70°, and the radius R of its circumcircle is 10 cm. Calculate the area of the triangle.

The standard formula for the area Δ of a triangle is $\Delta = \frac{1}{2}ab \sin C$.

You can find a and b from the full version of the sine formula.

Thus, from $2R = \dfrac{a}{\sin A} = \dfrac{b}{\sin B}$,

$$a = 2R \sin A = 20 \sin 50°$$

and
$$b = 2R \sin B = 20 \sin 60°.$$

Using these in the formula $\Delta = \frac{1}{2}ab \sin C$,

$$\Delta = \tfrac{1}{2} \times 20 \sin 50° \times 20 \sin 60° \times \sin 70°$$
$$= 200 \sin 50° \sin 60° \sin 70°.$$

The area of the triangle is $200 \sin 50° \sin 60° \sin 70°$ cm^2. ∎

The full version of the sine formula, $\dfrac{a}{\sin A} = \dfrac{b}{\sin B} = \dfrac{c}{\sin C} = 2R$, can often be used, together with the fact that the sum of the angles of a triangle is 180°, to prove other formulae concerned with a triangle.

As an example, here is a proof of the cosine formula using this method. This proof is certainly not the recommended proof of the cosine formula, but it does have the advantage that you do not need to consider the acute-angled and obtuse-angled triangles separately.

Example 13.3

Use the sine formula to prove the cosine formula

$$a^2 = b^2 + c^2 - 2bc \cos A.$$

Starting with the more complicated right-hand side, first substitute $b = 2R \sin B$ and $c = 2R \sin C$. Then manipulate the right-hand side, keeping symmetry as much as possible and using $A + B + C = 180$ judiciously, into a form that you can recognize as the left-hand side.

You use $A + B + C = 180°$ by substituting $B + C = 180 - A$ and then recognizing that $\sin(B + C)° = \sin(180 - A)° = \sin A°$ and that $\cos(B + C)° = \cos(180 - A)° = -\cos A°$.

$$\begin{aligned}
\text{RHS} &= b^2 + c^2 - 2bc \cos A° \\
&= 4R^2 \sin^2 B° + 4R^2 \sin^2 C° - 8R^2 \sin B° \sin C° \cos A° \\
&= 2R^2(2 \sin^2 B° + 2 \sin^2 C° - 4 \sin B° \sin C° \cos A°) \\
&= 2R^2((1 - \cos 2B°) + (1 - \cos 2C°) - 2 \cos A°(2 \sin B° \sin C°)) \\
&= 2R^2(2 - (\cos 2B° + \cos 2C°) - 2 \cos A°(2 \sin B° \sin C°)) \\
&= 2R^2(2 - (2 \cos(B + C)° \cos(B - C)°) - 2 \cos A°(2 \sin B° \sin C°)) \\
&= 2R^2(2 + 2 \cos A° \cos(B - C)° - 2 \cos A°(\cos(B - C)° - \cos(B + C)°)) \\
&= 2R^2(2 + 2 \cos A° \cos(B - C)° - 2 \cos A°(\cos(B - C)° + \cos A°)) \\
&= 2R^2(2 + 2 \cos A° \cos(B - C)° - 2 \cos A° \cos(B - C)° - 2 \cos^2 A°) \\
&= 2R^2(2 \sin^2 A°) \\
&= (2R \sin A°)^2 \\
&= a^2 = \text{LHS.}
\end{aligned}$$

Since RHS = LHS, the identity is true. ∎

Exercise 13.1

1 Calculate the radius of the circumcircle of the triangle ABC given that $a = 10\,\text{cm}$ and that angle $A = 30°$.

2 Find the exact radius of the circumcircle of a triangle with sides 2 cm, 3 cm and 4 cm, leaving square roots in your answer.

3 The area of a triangle ABC is $40\,\text{cm}^2$ and angles B and C are 50° and 70° respectively. Find the radius of the circumcircle.

4 Prove that $a = b \cos C + c \cos B$.

5 Let $s = \frac{1}{2}(a + b + c)$. Prove that $s = 4R \cos \frac{1}{2}A \cos \frac{1}{2}B \cos \frac{1}{2}C$.

13.2 The incircle

The incircle of a triangle ABC (see Figure 13.3) is the circle which touches each of the sides and lies inside the circle. The centre of the incircle will be denoted by I and its radius will be denoted by r.

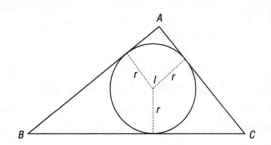

figure 13.3

To calculate the radius r of the incircle, let the perpendiculars from I on to the sides of the triangle ABC meet the sides at L, M and N. Join IA, IB and IC. See Figure 13.4.

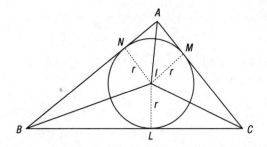

figure 13.4

Using the formula $\frac{1}{2}$ base $\times$ height, the area of the triangle BIC is

$$\frac{1}{2}ar.$$

Similarly, the area of the triangles CIA and AIB are

$$\frac{1}{2}br \text{ and } \frac{1}{2}cr.$$

Adding these you get Δ, the area of the triangle ABC. Therefore

$$\Delta = \frac{1}{2}ar + \frac{1}{2}br + \frac{1}{2}cr$$

$$= r \times \frac{a+b+c}{2}.$$

Denoting the expression $\frac{a + b + c}{2}$ by s, (for semi-perimeter), you find that

$$rs = \Delta. \qquad \qquad \textbf{Equation 2}$$

Equation 2 is used to derive results concerning the incircle of a triangle.

Example 13.4
The sides of a triangle are 4 cm, 5 cm and 6 cm. Calculate the radius of the incircle.

Let $a = 4$, $b = 5$ and $c = 6$. Use the cosine formula to calculate the cosine of the largest angle, C. Then find $\sin C$, and use this in the formula $\frac{1}{2} ab \sin C$ to find the area of the triangle. Then use Equation 2.

Using the cosine formula, $c^2 = a^2 + b^2 - 2ab \cos C$, gives

$$36 = 16 + 25 - 2 \times 4 \times 5 \cos C$$

which leads to $\qquad \cos C = \frac{1}{8}$.

Using $\sin^2 C = 1 - \cos^2 C$ shows that

$$\sin C = \sqrt{1 - \frac{1}{64}} = \sqrt{\frac{63}{64}} = \frac{3\sqrt{7}}{8}.$$

Using the formula $\frac{1}{2} ab \sin C$ for area, you obtain

$$\Delta = \frac{1}{2} \times 4 \times 5 \times \frac{3\sqrt{7}}{8} = \frac{15\sqrt{7}}{4}.$$

For this triangle $\qquad s = \frac{a + b + c}{2} = \frac{4 + 5 + 6}{2} = \frac{15}{2}.$

Therefore, using Equation 2, $rs = \Delta$,

$$r \times \frac{15}{2} = \frac{15\sqrt{7}}{4}$$

so $\qquad \qquad r = \frac{\sqrt{7}}{2}.$

The radius of the incircle is $\frac{1}{2}\sqrt{7}$ cm. ∎

13.3 The ecircles

The ecircles of a triangle ABC are the circles which touch each of the sides and lie outside the circle. There are three such circles. Figure 13.5 shows part of the circle opposite the vertex A.

The centre of the ecircle opposite A will be denoted by I_A and its radius will be denoted by r_A.

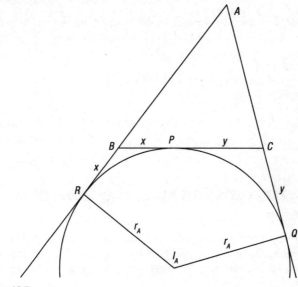

figure 13.5

To calculate r_A, let the points of contact of the ecircle with the sides of triangle ABC be P, Q and R.

Let $PB = x$ and $PC = y$. Then $RB = x$ and $QC = y$.

Then, from Figure 13.5, $x + y = a$ and as tangents from an external point, in this case A, are equal

$$c + x = b + y.$$

Solving the equations

$$x + y = a$$
$$x - y = b - c$$

simultaneously, gives

$$x = \frac{a + b - c}{2} = s - c \text{ and } y = \frac{a - b + c}{2} = s - b.$$

Notice also that $AR = c + x = c + (s - c) = s$ and that $AQ = s$.

Looking at areas, and noticing that the area Δ of triangle ABC is

$$\Delta = \text{area } ARI_AQ - \text{area } BRI_AP - \text{area } CQI_AP$$
$$= 2 \times \text{area } ARI_A - 2 \times \text{area } BRI_A - 2 \times \text{area } CQI_A$$
$$= 2 \times \tfrac{1}{2}r_A s - 2 \times \tfrac{1}{2}r_A(s-c) - 2 \times \tfrac{1}{2}r_A(s-b)$$
$$= r_A(s - (s-c) - (s-b))$$
$$= r_A(b + c - s)$$
$$= r_A(s - a).$$

Therefore $\qquad\qquad\qquad \Delta = r_A(s - a).$ **Equation 3**

There are related formulae for r_B and r_C.

13.4 Heron's formula: the area of a triangle

A number of the formulae in the previous section involved the semi-perimeter of a triangle, that is, s. There is a very famous formula, called Heron's formula, involving s for the area of a triangle. It is

$$\Delta = \sqrt{s(s-a)(s-b)(s-c)}.$$

The method of establishing Heron's formula comes from first using the cosine formula to calculate $\cos A$. The value of $\sin A$ is then calculated using $\sin^2 A = 1 - \cos^2 A$. Finally this value is substituted in the formula for area, $\Delta = \tfrac{1}{2}bc \sin A$.

Rearranging the cosine formula $a^2 = b^2 + c^2 - 2bc \cos A$ gives

$$\cos A = \frac{b^2 + c^2 - a^2}{2bc}.$$

Then using $\sin^2 A = 1 - \cos^2 A = (1 + \cos A)(1 - \cos A)$, you obtain

$$\sin^2 A = (1 + \cos A)(1 - \cos A)$$
$$= \left(1 + \frac{b^2 + c^2 - a^2}{2bc}\right)\left(1 - \frac{b^2 + c^2 - a^2}{2bc}\right)$$
$$= \frac{(2bc + b^2 + c^2 - a^2)(2bc - b^2 - c^2 + a^2)}{4b^2c^2}$$
$$= \frac{((b+c)^2 - a^2)(a^2 - (b-c)^2)}{4b^2c^2}$$

$$= \frac{(b + c + a)(b + c - a)(a + b - c)(a - b + c)}{4b^2c^2}$$

$$= \frac{2s(2(s - a))(2(s - c))(2(s - b))}{4b^2c^2}$$

$$= \frac{4s(s - a)(s - b)(s - c)}{b^2c^2}.$$

Therefore

$$\Delta = \tfrac{1}{2}bc \sin A$$

$$= \tfrac{1}{2}bc \times \frac{\sqrt{4s(s - a)(s - b)(s - c)}}{bc}$$

$$= \sqrt{s(s - a)(s - b)(s - c)}.$$

Therefore $\qquad \Delta = \sqrt{s(s - a)(s - b)(s - c)}.$ **Equation 4**

Equation 4 is called Heron's formula.

You can use the four equations established in this chapter, together with the technique used in Example 13.4, to establish most of the formulae you need.

Example 13.5

Find the radii of the three incircles of the triangle with sides of 4 cm, 5 cm and 6 cm.

Let $a = 4$, $b = 5$ and $c = 6$.

Then $s = \tfrac{1}{2}(a + b + c) = \tfrac{15}{2}$, $s - a = \tfrac{7}{2}$, $s - b = \tfrac{5}{2}$ and $s - c = \tfrac{3}{2}$.

In Equation 3, $\Delta = r_A(s - a)$, use Heron's formula, Equation 4, to find the area of the triangle.

Then $\qquad \Delta = \sqrt{\dfrac{15}{2} \times \dfrac{7}{2} \times \dfrac{5}{2} \times \dfrac{3}{2}} = \dfrac{15}{4}\sqrt{7}.$

Therefore, from $\Delta = r_A(s - a)$

$$\frac{15}{4}\sqrt{7} = \frac{7}{2}r_A$$

so $\qquad r_A = \dfrac{15\sqrt{7}}{4} \times \dfrac{2}{7} = \dfrac{15\sqrt{7}}{14}.$

Similarly $\qquad r_B = \dfrac{15\sqrt{7}}{4} \times \dfrac{2}{5} = \dfrac{3\sqrt{7}}{2}$

and $\qquad r_C = \dfrac{15\sqrt{7}}{4} \times \dfrac{2}{3} = \dfrac{5\sqrt{7}}{2}.$

The radii of the incircles are $\frac{15\sqrt{7}}{14}$ cm, $\frac{3\sqrt{7}}{2}$ cm and $\frac{5\sqrt{7}}{2}$ cm. ∎

Notice that the area of this triangle was found by an alternative method in Example 13.2.1.

Example 13.6

Prove that $\sin \frac{1}{2}A = \sqrt{\dfrac{(s-b)(s-c)}{bc}}$.

Use the cosine formula to find $\cos A$, and then use Equation 16 in Chapter 10, $\cos A = 1 - 2\sin^2 \frac{1}{2}A$, to find $\sin \frac{1}{2}A$.

Rearranging the cosine formula $a^2 = b^2 + c^2 - 2bc \cos A$ gives

$$\cos A = \frac{b^2 + c^2 - a^2}{2bc}.$$

Therefore

$$1 - 2\sin^2 \tfrac{1}{2}A = \frac{b^2 + c^2 - a^2}{2bc}$$

so

$$2\sin^2 \tfrac{1}{2}A = 1 - \frac{b^2 + c^2 - a^2}{2bc}.$$

Therefore

$$\begin{aligned}
\sin^2 \tfrac{1}{2}A &= \frac{1}{2}\left(1 - \frac{b^2 + c^2 - a^2}{2bc}\right) \\
&= \frac{2bc - b^2 - c^2 + a^2}{4bc} \\
&= \frac{a^2 - (b - c)^2}{4bc} \\
&= \frac{(a + (b - c))(a - (b - c))}{4bc} \\
&= \frac{(2(s - c))(2(s - b))}{4bc} \\
&= \frac{(s - b)(s - c)}{bc}.
\end{aligned}$$

Therefore

$$\sin \tfrac{1}{2}A = \sqrt{\frac{(s - b)(s - c)}{bc}}. \ ∎$$

Exercise 13.2

1 For the triangle with sides of length 2 cm, 3 cm and 4 cm, find the area and the radii of the incircle and the three ecircles.

2 Find the radius of the incircle of the triangle with sides 3 cm, 4 cm with an angle of 60° between them.

3 Prove that $rr_Ar_Br_C = \Delta^2$.

4 Prove that $r = 8R \sin \frac{1}{2}A \sin \frac{1}{2}B \sin \frac{1}{2}C$.

5 Prove that $r_C = 4R \cos \frac{1}{2}A \cos \frac{1}{2}B \sin \frac{1}{2}C$.

6 In Example 13.4.2 a formula for $\sin \frac{1}{2}A$ is derived. Use a similar method to derive a formula for $\cos \frac{1}{2}A$, and then use both of them to find a formula for $\tan \frac{1}{2}A$.

7 Prove the formula $\tan \frac{1}{2}(B - C) = \dfrac{b - c}{b + c} \cot \frac{1}{2}A$.

8 Prove that $r = (s - a) \tan \frac{1}{2}A$.

14

general solution of equations

In this chapter you will learn:
- how to find all the solutions of simple trigonometric equations
- general formulae for these solutions.

14.1 The equation sin θ = sin α

In earlier chapters you have always, when asked to solve an equation, been given an interval such as −180 to 180 or 0 to 360 in which to find the solutions. In this chapter the task will be to find a way of giving all solutions of a trigonometric equation, not just those solutions confined to a given interval.

For example, if you are given the equation $\sin\theta° = 0.6427...$ you can immediately look up the corresponding principal angle, in this case 40. You can therefore replace the equation $\sin\theta° = 0.6427...$ by the equation $\sin\theta° = \sin 40°$ and the problem of finding all solutions of $\sin\theta° = 0.6427...$ by finding all solutions of $\sin\theta° = \sin 40°$.

In Section 6.2 you saw that if 40 is a solution, then $180 - 40 = 140$ is also a solution, and that you can add any whole number multiple of 360 to give all solutions.

Thus all solutions of $\sin\theta° = \sin 40°$ are

$$..., -320, 40, 400, 760,...$$

and $$..., -220, 140, 500, 860,.... .$$

You can write these solutions in the form

$$40 + 360n \text{ and } 140 + 360n. \quad (n \text{ an integer})$$

Notice that these two formulae follow a pattern. Starting from 40 you can write them, in ascending order, as

$$0 \times 180 + 40$$
$$1 \times 180 - 40$$
$$2 \times 180 + 40$$
$$3 \times 180 - 40$$
$$\vdots$$

The pattern works backwards also, so all solutions fall into the pattern

$$\vdots$$
$$-2 \times 180 + 40$$
$$-1 \times 180 - 40$$
$$0 \times 180 + 40$$
$$1 \times 180 - 40$$
$$2 \times 180 + 40$$
$$3 \times 180 - 40$$
$$\vdots$$

You can use this pattern to write all the solutions in one formula as

$$180n + (-1)^n\,40$$

where $(-1)^n$ takes the value -1 when n is odd and $+1$ when n is even.

You can generalize this result further. If you solve for θ the equation $\sin\theta° = \sin\alpha°$ to give the solutions in terms of α you would obtain

$$\theta = 180n + (-1)^n\alpha \quad (n \text{ an integer}). \qquad \textbf{Equation 1}$$

Notice that in the formula $180n + (-1)^n\alpha$ there is no reason why α should be an acute angle.

If the original angle had been 140, the formula

$$\theta = 180n + (-1)^n 140$$

where n is an integer gives the solutions

$$\ldots, -360 + 140, -180 - 140, 0 + 140, 180 - 140, 360 + 140, \ldots$$

or $\qquad\qquad \ldots, -220, -320, 140, 40, 500, \ldots$.

You can see that these solutions are the same solutions as those in the middle of page 139, but they are in a different order.

Therefore the formula $180n + (-1)^n\alpha$ where n is an integer gives you all solutions of the equation $\sin\theta° = \sin\alpha°$ provided you have one solution α.

In radians, the same formula gives

$$\theta = n\pi + (-1)^n\alpha \quad (n \text{ an integer}) \qquad \textbf{Equation 2}$$

as the solution of the equation $\sin\theta = \sin\alpha$.

Example 14.1
Find all the solutions in degrees of the equation $\sin\theta° = -\frac{1}{2}$.

The principal solution of the equation $\sin\theta° = -\frac{1}{2}$ is -30.

Using Equation 1, all solutions are $180n + (-1)^n(-30)$ or alternatively, $180n - (-1)^n 30$ for integer n. ∎

Example 14.2
Find all solutions in radians of the equation $\sin(2\theta + \frac{1}{6}\pi) = \frac{1}{2}\sqrt{3}$.

Let $y = 2\theta + \frac{1}{6}\pi$. Then the equation becomes $\sin y = \frac{1}{2}\sqrt{3}$.

The principal angle is $\frac{1}{3}\pi$. So the solution of the equation is

$$y = n\pi + (-1)^n \tfrac{1}{3}\pi.$$

Therefore $\qquad 2\theta + \tfrac{1}{6}\pi = n\pi + (-1)^n \tfrac{1}{3}\pi$

so $\qquad \theta = \tfrac{1}{2}n\pi + (-1)^n \tfrac{1}{6}\pi - \tfrac{1}{12}\pi$ (n an integer). ∎

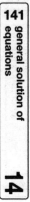
14.2 The equation $\cos\theta = \cos\alpha$

It is convenient to consider the equation $\cos\theta° = 0.7660...$, that is, the equation $\cos\theta° = \cos 40°$.

Using the method of Section 6.3, the solutions of this equation are

$$..., -320, 40, 400, 760,...$$

and $\qquad ..., -400, -40, 320, 680,... .$

You can write these solutions in the form

$$40 + 360n \text{ and } -40 + 360n \qquad (n \text{ an integer}).$$

The pattern for these solutions is easier to follow than that for the sine function. It is

$$\theta = 360n \pm 40 \quad (n \text{ an integer}).$$

You can generalize this further, as in the case of the sine function. The solution of the equation $\cos\theta° = \cos\alpha°$ is

$$\theta = 360n \pm \alpha \text{ (n an integer).} \qquad \textbf{Equation 3}$$

In radians, the same formula gives

$$\theta = 2n\pi \pm \alpha \text{ (n an integer)} \qquad \textbf{Equation 4}$$

as the solution of the equation $\cos\theta = \cos\alpha$.

14.3 The equation $\tan\theta = \tan\alpha$

Consider the equation $\tan\theta° = \tan 40°$.

Using the method of Section 6.4, the solutions of this equation are

$$..., -320, -140, 40, 220, 400, 580, 760,... .$$

An alternative form is

$$40 + 180n \text{ (n an integer).}$$

The solution of the equation $\tan\theta° = \tan\alpha°$ is

$$\theta = 180n + \alpha \text{ (n an integer).} \qquad \textbf{Equation 5}$$

In radians, the same formula gives

$$n\pi + \alpha \text{ (n an integer)} \qquad \textbf{Equation 6}$$

as the solution of the equation $\tan\theta = \tan\alpha$.

14.4 Summary of results

Here is a summary of the results of the previous sections.

The general solution in degrees of the equation $\sin\theta° = \sin\alpha°$ is

$$\theta = 180n + (-1)^n\alpha \text{ (n an integer).} \qquad \textbf{Equation 1}$$

The general solution in radians of the equation $\sin\theta = \sin\alpha$ is

$$\theta = n\pi + (-1)^n\alpha \text{ (n an integer).} \qquad \textbf{Equation 2}$$

The general solution in degrees of the equation $\cos\theta° = \cos\alpha°$ is

$$\theta = 360n \pm \alpha \text{ (n an integer).} \qquad \textbf{Equation 3}$$

The general solution in radians of the equation $\cos\theta = \cos\alpha$ is

$$\theta = 2n\pi \pm \alpha \text{ (n an integer).} \qquad \textbf{Equation 4}$$

The general solution in degrees of the equation $\tan\theta° = \tan\alpha°$ is

$$\theta = 180n + \alpha \text{ (n an integer).} \qquad \textbf{Equation 5}$$

The general solution in radians of the equation $\tan\theta = \tan\alpha$ is

$$\theta = n\pi + \alpha \text{ (n an integer).} \qquad \textbf{Equation 6}$$

Example 14.3

Find the general solution in degrees of the equation $\cos\theta° = -\frac{1}{2}\sqrt{3}$.

The principal angle is 150, so the general solution is

$$\theta = 360n \pm 150 \text{ (n an integer).} \blacksquare$$

Example 14.4

Find the general solution in radians of the equation $\tan 2x = -\sqrt{3}$.

Let $2x = y$. The principal angle for equation $\tan y = -\sqrt{3}$ $-\frac{1}{3}\pi$, so the general solution for y is

$$y = n\pi \pm (-\tfrac{1}{3}\pi) \text{ (}n \text{ an integer)}.$$

This is the same as $y = n\pi \pm \tfrac{1}{3}\pi$ (n an integer).

Therefore, as $x = \tfrac{1}{2}y$, $x = \tfrac{1}{2}n\pi \pm \tfrac{1}{6}\pi$ (n an integer). ∎

Example 14.5

Solve in radians the equation $\cos 2\theta = \cos \theta$.

Using Equation 4, $2\theta = 2n\pi \pm \theta$ (n an integer).

Taking the positive sign gives

$$\theta = 2n\pi \text{ (}n \text{ an integer)},$$

and the negative sign $3\theta = 2n\pi$ (n an integer).

Therefore the complete solution is $\theta = \tfrac{2}{3}n\pi$ (n an integer). ∎

Notice that the solution given, $\theta = \tfrac{2}{3}n\pi$ (n an integer), includes, when n is a multiple of three, the solution obtained by taking the positive sign.

Example 14.6

Solve in radians the equation $\sin 3\theta = \sin \theta$.

Using Equation 1,

$$3\theta = n\pi \pm (-1)^n \theta \text{ (}n \text{ an integer)}.$$

Taking n even, and writing $n = 2m$ gives

$$3\theta = 2m\pi + \theta, \text{ that is, } \theta = m\pi.$$

Taking n odd, and writing $n = 2m + 1$ gives

$$3\theta = (2m + 1)\pi - \theta, \text{ that is, } 4\theta = (2m + 1)\pi.$$

Thus the complete solution is

either $\theta = m\pi$ or $\theta = \tfrac{1}{4}(2m + 1)\pi$ where m is an integer. ∎

Example 14.7

Solve the equation $\cos 2\theta° = \sin \theta°$ giving all solutions in degrees.

Use the fact that $\sin \theta° = \cos(90 - \theta)°$ to rewrite the equation as

$$\cos 2\theta° = \cos(90 - \theta)°.$$

Using Equation 3,

$$2\theta = 360n \pm (90 - \theta) \ (n \text{ an integer}).$$

Taking the positive sign gives

$$3\theta = 360n + 90 \text{ or } \theta = 120n + 30,$$

and the negative sign

$$\theta = 360n - 90.$$

Thus the complete solution is

either $\theta = 120n + 30$ or $\theta = 360n - 90$ where n is an integer. ∎

Compare this solution with that for Example 10.8.

Exercise 14

In questions 1 to 10 find the general solution in degrees of the given equation.

1 $\sin \theta° = \frac{1}{2}\sqrt{3}$ **2** $\cos \theta° = \frac{1}{2}$

3 $\tan \theta° = -1$ **4** $\sin(2\theta - 30)° = -\frac{1}{2}$

5 $\cos 3\theta° = -1$ **6** $\tan(3\theta + 60)° = -\frac{1}{3}\sqrt{3}$

7 $\tan 2\theta° = \tan \theta°$ **8** $\cos 3\theta° = \cos 2\theta°$

9 $\sin 3\theta° = \sin \theta°$ **10** $\sin 3\theta° = \cos \theta°$

In questions 11 to 20 give the general solution in radians of the given equation.

11 $\sin 2\theta = -\frac{1}{2}$ **12** $\cos 3\theta = 0$

13 $\tan(2\theta + \frac{1}{2}\pi) = 0$ **14** $\sin(3\theta + \frac{1}{6}\pi) = \frac{1}{2}$

15 $\cos(2\theta + \frac{1}{2}\pi) = 0$ **16** $\tan(\frac{1}{2}\pi - 2\theta) = -1$

17 $\tan 3\theta = \cot(-\theta)$ **18** $\cos 3\theta = \sin 2\theta$

19 $\cos 3\theta = \sin(\frac{1}{2}\pi - \theta)$ **20** $\sin 2\theta = -\sin \theta$

angle of depression The angle of depression of B from A, where B is below A, is the angle θ that the line AB makes with the horizontal. See Figure 15.1.

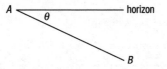

figure 15.1

angle of elevation The angle of elevation of B from A, where B is above A, is the angle θ that the line AB makes with the horizontal. See Figure 15.2.

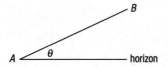

figure 15.2

bearing The bearing θ of B from A is the angle in degrees between north and the line AB, measured clockwise from the north. See Figure 15.3.

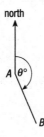

figure 15.3

corresponding angles When two parallel lines, *AB* and *CD*, are traversed by a line *XY*, the marked angles in Figure 15.4 are called corresponding angles, and are equal.

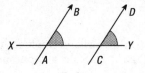

figure 15.4

isosceles triangle An isosceles triangle is a triangle with two equal sides. In Figure 15.5, *AB = AC*.

figure 15.5

kite A kite is a quadrilateral which has one line of reflective symmetry. In Figure 15.6, *ABCD* is a kite with *AC* as its axis of symmetry.

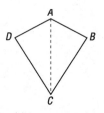

figure 15.6

periodic, period A periodic function $f(x)$ is a function with the property that there exists a number c such that $f(x) = f(x + c)$ for all values of x. The smallest such value of c is called the period of the function.

Pythagoras's theorem Pythagoras's theorem states that in a right-angled triangle with sides a, b and c, where c is the hypotenuse, $c^2 = a^2 + b^2$. See Figure 15.7.

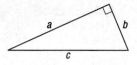

figure 15.7

rhombus A rhombus is a quadrilateral which has all four sides equal in length. See Figure 15.8. In a rhombus, the diagonals cut at right angles, and bisect each other.

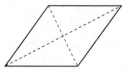

figure 15.8

similar triangles Two triangles which have equal angles are similar. The sides of similar triangles are proportional to each other. In Figure 15.9, triangles ABC and XYZ are similar, and $\dfrac{BC}{YZ} = \dfrac{CA}{ZX} = \dfrac{AB}{XY}$.

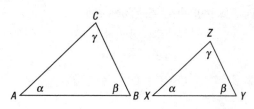

figure 15.9

slant height, of a cone The slant height of the cone is the length of the sloping edge AB in Figure 15.10.

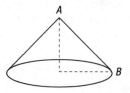

figure 15.10

Relations between the ratios

$$\tan\theta = \frac{\sin\theta}{\cos\theta}$$

Pythagoras's equation

$$\sin^2\theta + \cos^2\theta = 1$$
$$1 + \tan^2\theta = \sec^2\theta$$
$$1 + \cot^2\theta = \operatorname{cosec}^2\theta$$

Relations for solving equations

$$\sin\theta° = \sin(180 - \theta)°$$
$$\cos\theta = \cos(-\theta)$$
$$\tan\theta° = \tan(180 + \theta)°$$

Solution of triangles

$$\Delta = \tfrac{1}{2}bc\sin A = \tfrac{1}{2}ca\sin B = \tfrac{1}{2}ab\sin C$$
$$\frac{\sin A}{a} = \frac{\sin B}{b} = \frac{\sin C}{c}$$
$$a^2 = b^2 + c^2 - 2bc\cos A$$
$$b^2 = c^2 + a^2 - 2ca\cos B$$
$$c^2 = a^2 + b^2 - 2ab\cos C$$

Compound angles

$$\sin(A + B) = \sin A\cos B + \cos A\sin B$$
$$\sin(A - B) = \sin A\cos B - \cos A\sin B$$

$$\cos(A + B) = \cos A \cos B - \sin A \sin B$$

$$\cos(A - B) = \cos A \cos B + \sin A \sin B$$

$$\tan(A + B) = \frac{\tan A + \tan B}{1 - \tan A \tan B}$$

$$\tan(A - B) = \frac{\tan A - \tan B}{1 + \tan A \tan B}$$

Multiple angles

$$\sin 2A = 2 \sin A \cos A$$

$$\cos 2A = \cos^2 A - \sin^2 A$$

$$\cos 2A = 2 \cos^2 A - 1$$

$$\cos 2A = 1 - 2 \sin^2 A$$

$$\tan 2A = \frac{2 \tan A}{1 - \tan^2 A}$$

Factor formulae

$$2 \sin A \cos B = \sin(A + B) + \sin(A - B)$$

$$2 \cos A \sin B = \sin(A + B) - \sin(A - B)$$

$$2 \cos A \cos B = \cos(A + B) + \cos(A - B)$$

$$2 \sin A \sin B = \cos(A - B) - \cos(A + B)$$

$$\sin C + \sin D = 2 \sin \frac{C + D}{2} \cos \frac{C - D}{2}$$

$$\sin C - \sin D = 2 \cos \frac{C + D}{2} \sin \frac{C - D}{2}$$

$$\cos C + \cos D = 2 \cos \frac{C + D}{2} \cos \frac{C - D}{2}$$

$$\cos C - \cos D = 2 \sin \frac{C + D}{2} \sin \frac{D - C}{2}$$

General solution of equations

The general solution in degrees of the equation $\sin \theta° = \sin \alpha°$ is

$$\theta = 180n + (-1)^n \alpha \ (n \text{ an integer}).$$

The general solution in radians of the equation $\sin \theta = \sin \alpha$ is

$$\theta = n\pi + (-1)^n \alpha \ (n \text{ an integer}).$$

The general solution in degrees of the equation $\cos\theta° = \cos\alpha°$ is
$$\theta = \sin 360n \pm \alpha \; (n \text{ an integer}).$$
The general solution in radians of the equation $\cos\theta = \cos\alpha$ is
$$\theta = 2n\pi \pm \alpha \; (n \text{ an integer}).$$
The general solution in degrees of the equation $\tan\theta° = \tan\alpha°$ is
$$\theta = 180n + \alpha \; (n \text{ an integer}).$$
The general solution in radians of the equation $\tan\theta = \tan\alpha$ is
$$\theta = n\pi + \alpha \; (n \text{ an integer}).$$

Radians
$$\pi \text{ rad} = 180 \text{ degrees}$$

Length of arc
$$s = r\theta$$

Area of circular sector
$$A = \tfrac{1}{2}r^2\theta$$

Radius of circumcircle
$$2R = \frac{a}{\sin A} = \frac{b}{\sin B} = \frac{c}{\sin C}$$

Radius of incircle
$$rs = \Delta$$
where $s = \tfrac{1}{2}(a + b + c)$.

Radius of ecircle
$$\Delta = r_A(s - a) = r_B(s - b) = r_C(s - c)$$
where $s = \tfrac{1}{2}(a + b + c)$.

Heron's formula for area of triangle
$$\Delta = \sqrt{s(s - a)(s - b)(s - c)}$$
where $s = \tfrac{1}{2}(a + b + c)$.

answers

Answers involving lengths are given either correct to three significant figures or to three decimal places, whichever seems more appropriate. Answers for angles are usually given correct to two decimal places.

Exercise 2.1 (page 9)

1	0.364	2	0.577
3	5729.578	4	0.852
5	1.881	6	0.009
7	18.88°	8	63.43°
9	80.72°	10	0.01°
11	45°	12	60°

Exercise 2.2 (page 13)

1	8.36 m	2	038.66°
3	4.69 miles	4	19.54 m
5	126 m	6	21.3 m
7	53.01°, 36.99°	8	144 m
9	75.64°		

Exercise 2.3 (page 17)

1	68.20°	2	3.73 cm
3	2.51 cm	4	48.81°
5	11.59 cm	6	3.46 cm
7	8.06 cm	8	24.78°
9	63.86°	10	3.60 cm

Exercise 3.1 (page 24)

1	66.42°	2	5.47 cm
3	2.25 cm	4	28.96°
5	3.37 cm	6	1.73 cm
7	6.85 cm	8	27.49°
9	60.61°	10	2.68 cm
11	0.994, 0.111	12	5.14 cm, 3.06 cm

13 36.87°, 53.13°   14 38.43°
15 9.32 km   16 4.82 cm
17 2.87 cm   18 7.19 km
19 38.42°, 51.58°, 7.40 cm

Exercise 3.2 (page 29)

1 73.30°   2 6.66 cm
3 5.79 cm   4 48.81°
5 9.44 cm   6 2.1 cm
7 10.03 cm   8 67.38°
9 19.02°   10 5.38 cm
11 35.02°, 2.86 cm   12 44.20°
13 55.5 mm, 72.7 mm   14 30.50°, 59.50°
15 2.66 cm, 1.86 cm, 3.88 cm 16 44.12°, 389.8 mm
17 69.51°   18 14.7 km, 10.3 km
19 0.681 cm   20 $\frac{1}{2}x\sqrt{3}$ m, $\frac{1}{2}\sqrt{3}$, $\frac{1}{2}$
21 2.60 cm, 2.34 cm   22 3.59°
23 10.17 km, 11.70 km   24 328.17°, 17.1 km

Exercise 3.3 (page 33)

1 10.05 cm   2 4.61 cm
3 145.42°   4 4.33 cm
5 10.91 km, 053.19°   6 85.06°
7 44.80 cm   8 8.58 cm
9 25.24 cm   10 13.20 cm

Exercise 4 (page 43)

1 68.20°, 60.50°   2 73.57°, 78.22°
3 3.68 m   4 97.08°
5 35.26°   6 12.72°, 15.76°
7 54.74°   8 5.20 cm, 50.77°
9 56.31°   10 35.26°

Exercise 5.1 (page 50)

1 0.766   2 −0.766
3 −0.940   4 0.985
5 −0.342   6 0.174
7 0   8 0
9 3   10 1
11 1   12 1
13 2   14 4
15 1   16 0
17 0   18 −1
19 −1   20 1

Exercise 5.2 (page 52)

1	−1.732	**2**	−0.577
3	0.364	**4**	−5.671
6	0.213	**7**	−0.176

Exercise 6.1 (page 57)

1	17.46, 162.54	**2**	26.74, 153.26
3	0, 180, 360	**4**	90
5	270	**6**	185.74, 354.26
7	206.74, 333.26	**8**	210, 330
9	−171.37, −8.63	**10**	−150, −30
11	−180, 0, 180	**12**	90
13	−90	**14**	64.16, 115.84
15	−115.84, −64.16	**16**	−130.00, −50.00
17	15, 75, 195, 255	**18**	13.37, 72.63, 193.37, 256.63
19	0, 60, 120, 180, 240, 300, 360	**20**	135, 315
21	60, 300	**22**	180
23	90, 210, 330	**24**	20.91, 69.09, 200.91, 249.09
25	1.62 hours	**26**	About 122 days

Exercise 6.2 (page 61)

1	109.47, 250.53	**2**	63.43, 243.43
3	41.41, 318.59	**4**	153.43, 333.43
5	30, 150, 210, 330	**6**	22.5, 112.5, 202.5, 292.5
7	203.07	**8**	143.18

9 70.00, 110.00, 250.00, 290.00
10 87.14, 177.14, 267.14, 357.14
11 −126.27, −53.73, 53.73, 126.27
12 −103.28, −13.28, 76.72, 166.72
13 −168.21, −101.79, 11.79, 78.21
14 −90, 90
15 −90, 30, 150
16 −45
17 4.14 pm and 7.46 pm

Exercise 7.1 (page 69)

These answers are given in alphabetical order.

1 15.82 cm, 14.73 cm, 94.22 cm^2
2 20.29 cm, 30.36 cm, 152.22 cm^2
3 7.18 mm, 6.50 mm, 18.32 mm^2
4 5.59 cm, 7.88 cm, 22.00 cm^2
5 23.06 cm, 17.32 cm, 192.88 cm^2
6 $C = 28.93°$, $A = 126.07°$, $a = 58.15$ cm or
$C = 151.07°$, $A = 3.93°$, $a = 4.93$ cm

7 $C = 51.31°$, $A = 88.69°$, $a = 109.26$ cm or
  $C = 128.69°$, $A = 11.31°$, $a = 21.43$ cm
8 $A = 61.28°$, $B = 52.72°$, $b = 87.10$ cm
9 $A = 35.00°$, $B = 115.55°$, $b = 143.13$ cm or
  $A = 145.00°$, $B = 5.55°$, $b = 15.35$ cm

Exercise 7.2 (page 72)

1	37.77 cm	**2**	5.30 cm
3	54.73 cm	**4**	25.9 cm
5	2.11 cm	**6**	7.97 cm
7	28.96°, 46.57°, 104.47°	**8**	40.11°, 57.90°, 81.99°
9	62.19°, 44.44°, 73.37°	**10**	28.91°, 31.99°, 119.10°
11	106.23°	**12**	43.84°

13 You get an error message from the calculator because the cosine of the largest angle is –1.029, which is impossible. The problem arises because the triangle is impossible to draw, as the longest side is longer than the sum of the other two sides.

14	5.93 km	**15**	52.01°, 88.05°, 39.93°
16	45.17°, 59.60°, 7.25 cm	**17**	56.09°
18	16.35 m, 13.62 m	**19**	41.04°

20 Two triangles. 66.82°, 63.18° and 29.13 cm
  16.82°, 113.18° and 9.44° cm

21	98.34 m	**22**	5.71 m, 6.08 m
23	3.09 mm	**24**	7.98 cm, 26.32° and 29.93°
25	3.81 cm, 4.20 cm, 7.81 cm^2	**26**	4.51 hours
27	4.41 km	**28**	0.305 m^2
29	49.46°, 58.75°		

Exercise 7.3 (page 80)

1	15.2 m	**2**	546 m
3	276 m	**4**	192 m
5	889 m	**6**	1.26 km
7	3700 m	**8**	2.23 km
9	2.88 km	**10**	2.17 km
11	500 m	**12**	3.64 km, 315°, 5.15 km
13	72.9 m, 51.1 m	**14**	1.246 km
15	189 m	**16**	63.7 m
17	3470 m, 7270 m		

Exercise 8 (page 88)

1	60	**2**	15
3	270	**4**	120
5	135	**6**	720
7	0.588	**8**	0.924
9	0.309	**10**	0.383

11 0.966	**12** 0.5
13 13.41	**14** $\frac{1}{12}\pi$
15 $\frac{2}{5}\pi$	**16** $\frac{11}{30}\pi$
17 $\frac{7}{12}\pi$	**18** 4.75
19 1.12 cm	**20** 6.72 cm^2
21 1.6	**22** 0.611 rad, 35.01°
23 $\frac{1}{4}\pi$, $\frac{1}{3}\pi$ and $\frac{5}{12}\pi$	**24** 23.18 cm^2 and 55.36 cm^2
25 2.90 cm^2	**26** 3.03 cm^2

Exercise 9 (page 95)

1 −0.5735

2 ±3.180

3 ±1.077

4 $\dfrac{k}{\sqrt{k^2-1}}$

5 $-\sqrt{1+t^2}, -\dfrac{1}{\sqrt{1+t^2}}, \dfrac{t}{\sqrt{1+t^2}}$

6 $\dfrac{1}{\sqrt{s^2-1}}, \dfrac{\sqrt{s^2-1}}{s}$

7 −120, −60, 60, 120

8 −135, −45, 45, 135

9 −180, 0, 30, 150, 180

10 −180, 0, 180

11 38.17, 141.83

12 −116.57, 63.43

13 −153.43, 26.57

14 −75.52, 0, 75.52

15 −60, 30, 60, 150

Exercise 10.1 (page 102)

1 0.663, −0.749

2 $\dfrac{\sqrt{6}-\sqrt{2}}{4}, \dfrac{\sqrt{6}-\sqrt{2}}{4}$

3 Expanding gives $\sin(90 - \theta)° = \sin 90° \cos \theta° - \cos 90° \sin \theta°$
$= 1 \times \cos \theta° - 0 \times \sin \theta° = \cos \theta°$

4 −0.997

5 0.894, 1.999

6 $\dfrac{3+\sqrt{3}}{3-\sqrt{3}}$ or $\dfrac{\sqrt{3}+1}{\sqrt{3}-1}$ or $2+\sqrt{3}$

7 $\frac{40}{13} \approx 3.077, \frac{20}{37} \approx 0.541$

8 Using Equation 6, $\tan(180 - \theta)° = \dfrac{\tan 180° - \tan \theta°}{1 + \tan 180° \tan \theta°}$

$= \dfrac{0 - \tan \theta°}{1 + 0 \times \tan \theta°} = -\tan \theta°$

9 $\sin 34°$

10 $\cos 61°$

11 $\tan 68°$

12 $\tan 39°$

Exercise 10.2 (page 105)

1 $\frac{24}{25}, \frac{7}{25}, \frac{24}{7}$

2 $-\frac{24}{25}, \frac{7}{25}, -\frac{24}{7}$

3 0.484, 0.875, 0.553

4 1, 0

5 0.992, −0.129

6 −0.992, −0.129

7 $\sin 72°, \cos 72°$

8 0.5, −0.5

9 Use Equation 16, $\cos\theta = 1 - 2\sin^2\frac{1}{2}\theta$, to get $\sin^2\frac{1}{2}\theta = \dfrac{1-\cos\theta}{2}$ and the result follows. Use Equation 15, $\cos\theta = 2\cos^2\frac{1}{2}\theta - 1$, to get $\cos^2\frac{1}{2}\theta = \dfrac{1-\cos\theta}{2}$ and the result follows.

10 $\pm\frac{1}{2}, \pm\frac{1}{2}\sqrt{3}$ **11** ±0.6

12 $\tan 20°$

Exercise 10.3 (page 108)

In the solutions to the identities, the reason for each step is given by placing an Equation number from Chapter 10 in bold-faced type in brackets. Where no reason is given, either an algebraic simplification or the use of $\sin^2 A + \cos^2 A = 1$ is involved.

1 LHS $= \sin(A + B) + \sin(A - B)$

$= \sin A\cos B + \cos A\sin B + \sin A\cos B - \cos A\sin B$ **(1 & 2)**

$= 2\sin A\cos B =$ RHS

2 LHS $= \cos(A + B) + \cos(A - B)$

$= \cos A\cos B - \sin A\sin B + \cos A\cos B + \sin A\sin B$ **(3 & 4)**

$= 2\cos A\cos B =$ RHS

3 LHS $= \dfrac{\cos 2A}{\cos A + \sin A}$

$= \dfrac{\cos^2 A - \sin^2 A}{\cos A + \sin A}$ **(9)**

$= \dfrac{(\cos A + \sin A)(\cos A - \sin A)}{\cos A + \sin A}$

$= \cos A - \sin A =$ RHS

4 LHS $= \sin 3A = \sin(2A + A)$

$= \sin 2A\cos A + \cos 2A\sin A$ **(1)**

$= 2\sin A\cos A \times \cos A + (1 - 2\sin^2 A) \times \sin A$ **(7 & 11)**

$= 2\sin A\cos^2 A + \sin A - 2\sin^3 A$

$= 2\sin A(1 - \sin^2 A) + \sin A - 2\sin^3 A$

$= 3\sin A - 4\sin^3 A =$ RHS

5 LHS $= \dfrac{\sin A}{\cos A} + \dfrac{\cos A}{\sin A}$

$= \dfrac{\sin^2 A + \cos^2 A}{\sin A\cos A}$

$= \dfrac{1}{\sin A\cos A} = \dfrac{2}{2\sin A\cos A}$

$= \dfrac{2}{\sin 2A}$ **(7)** $=$ RHS

6 LHS $= \dfrac{1}{\sin\theta} + \dfrac{\cos\theta}{\sin\theta}$

$= \dfrac{1+\cos\theta}{\sin\theta}$

$= \dfrac{2\cos^2\frac{1}{2}\theta}{2\sin\frac{1}{2}\theta\cos\frac{1}{2}\theta}$ **(12 & 8)**

$= \dfrac{\cos\frac{1}{2}\theta}{\sin\frac{1}{2}\theta} = $ RHS

7 LHS $= \dfrac{\sin\theta}{\sin\phi} + \dfrac{\cos\theta}{\cos\phi}$

$= \dfrac{\sin\theta\cos\phi + \sin\phi\cos\theta}{\sin\phi\cos\phi}$

$= \dfrac{\sin(\theta+\phi)}{\frac{1}{2}\times 2\sin\phi\cos\phi}$ **(1)**

$= \dfrac{2\sin(\theta+\phi)}{\sin 2\phi}$ **(7)** $= $ RHS

Exercise 10.4 (page 109)

1 −90, 30, 90, 150
2 −150, −30, 30, 150
3 −148.28, −58.28, 31.72, 122.72
4 −165, −105, 15, 75
5 −157.5, −67.5, 22.5, 112.5
6 −150, −30, 30, 150
7 −120, 0, 120
8 0, ±60, ±120, ±180
9 ±30, ±90, ±150
10 −157.5, −67.5, 22.5, 112.5

Exercise 11.1 (page 114)

1 $\sqrt{2}\sin(x+45)°$

2 $13\sin(x+67.38)°$

3 $\sqrt{29}\sin(x+21.80)°$

4 $\sqrt{2}\sin(x+135)°$

5 $\sqrt{10}\sin(x+288.43)°$

6 $\sqrt{10}\sin(x+108.43)°$

7 $\sqrt{2}\cos(x+\frac{7}{4}\pi)$

8 $\sqrt{2}\cos(x-\frac{1}{4}\pi)$

9 $\sqrt{2}\sin(x-\frac{7}{4}\pi)$

10 Both functions have the form $R\cos(x+\alpha)$ with $R=\sqrt{50}$. This means, see Section 11.3, that in both cases the maximum value is $\sqrt{50}$.

Exercise 11.2 (page 117)

1 0, 90, 360
2 90, 330
3 40.21, 252.41
4 45.24, 180
5 159.24, 283.13
6 45, 90, 225, 270
7 −180, −90, 180
8 −180, 60, 180
9 −69.20, 32.33
10 −157.38, −90
11 289.59, 349.20
12 −135, −75, 45, 105

13 $\sqrt{5}$ at $x = 116.57$, $-\sqrt{5}$ at $x = -63.43$

14 5 at $x = -53.13$, $- 5$ at $x = 126.87$

15 2 at $x = -15$ and 165, $- 2$ at $x = -105$ and 75

16 $\sqrt{2}$ at $x = -22.5$ and 157.5, $-\sqrt{2}$ at $x = -112.5$ and 67.5

17 7 at $x = 36.87$, $- 3$ at $x = -144.13$

18 4.732 at $x = -17.63$ and 162.37, 1.268 at $x = -107.63$ and 72.37

Exercise 12.1 (page121)

1 $\frac{1}{2}(\sin 4\theta + \sin 2\theta)$
2 $\frac{1}{2}(\sin 80° - \sin 10°)$
3 $\frac{1}{2}(\cos 80° + \sin 20°)$
4 $\frac{1}{2}(\sin 8\theta - \sin 2\theta)$
5 $\frac{1}{2}(\cos 3(C + D) + \cos(C - D))$
6 $\frac{1}{2}(\sin 90° - \sin 30°) = \frac{1}{4}$
7 $\cos 2A - \cos 4A$
8 $\frac{1}{2}(\sin 6C - \sin 10D)$
9 $2 \cos C \sin D = \sin(C + D) - \sin(C - D)$
10 $2 \cos C \cos B = \cos(C + B) + \cos(C - B)$

Exercise 12.2 (page125)

1 $2 \sin 3A \cos A$
2 $2 \cos 3A \sin 2A$
3 $-2 \sin 3\theta \sin \theta$
4 $2 \sin 3A \sin 2A$
5 $2 \cos 41° \cos 6°$
6 $2 \cos 36° \sin 13°$
7 $\cot 15°$
8 $\tan \frac{1}{2}(\alpha + \beta)$
9 $\tan \theta$
10 $\tan 2\theta$
11 0, 120, 240, 360
12 0, 120, 180, 240, 360
13 0, 90, 120, 180, 240, 270, 360
14 45, 135, 180, 225, 315
15 LHS $= \sin A + \sin B + \sin C$

$= (\sin A + \sin B) + \sin(A + B)$

$= 2 \sin \frac{1}{2}(A + B) \cos \frac{1}{2}(A - B) + 2 \sin \frac{1}{2}(A + B) \cos \frac{1}{2}(A + B)$

$= 2 \sin \frac{1}{2}(A + B)(\cos \frac{1}{2}(A - B) + \cos \frac{1}{2}(A + B))$

$= 2 \cos \frac{1}{2}C(2 \cos \frac{1}{2}A \cos \frac{1}{2}B)$

$= 4 \cos \frac{1}{2}A \cos \frac{1}{2}B \cos \frac{1}{2}C = $ RHS

16 LHS $= \sin A + \sin B - \sin C$

$= (\sin A + \sin B) - \sin(A + B)$

$= 2 \sin\frac{1}{2}(A + B) \cos\frac{1}{2}(A - B) - 2 \sin\frac{1}{2}(A + B) \cos\frac{1}{2}(A + B)$

$= 2 \sin\frac{1}{2}(A + B)(\cos\frac{1}{2}(A - B) - \cos\frac{1}{2}(A + B))$

$= 2 \cos\frac{1}{2}C(2 \sin\frac{1}{2}A \sin\frac{1}{2}B)$

$= 4 \sin\frac{1}{2}A \sin\frac{1}{2}B \cos\frac{1}{2}C = $ RHS

17 LHS $= \cos A + \cos B + \cos C - 1$

$= (\cos A + \cos B) + (\cos C - 1)$

$= 2 \cos\frac{1}{2}(A + B) \cos\frac{1}{2}(A - B) - 2 \sin^2\frac{1}{2}C$

$= 2 \sin\frac{1}{2}C (\cos\frac{1}{2}(A - B) - \sin\frac{1}{2}C)$

$= 2 \sin\frac{1}{2}C (\cos\frac{1}{2}(A - B) - \cos\frac{1}{2}(A + B))$

$= 2 \sin\frac{1}{2}C (2 \sin\frac{1}{2}A \sin\frac{1}{2}B)$

$= 4 \sin\frac{1}{2}A \sin\frac{1}{2}B \sin\frac{1}{2}C = $ RHS

Exercise 13.1 (page130)

1 10 cm

2 $\dfrac{8}{\sqrt{15}}$ cm

3 $\sqrt{\dfrac{20}{\sin 50° \sin 60° \sin 70°}}$ cm

4 RHS $= b \cos C + c \cos B$

$= 2R \sin B \cos C + 2R \sin C \cos B$

$= 2R(\sin B \cos C + \sin C \cos B)$

$= 2R(\sin(B + C))$

$= 2R \sin A = a = $ LHS

5 LHS $= s$

$= \frac{1}{2}(a + b + c)$

$= R(\sin A + \sin B + \sin C)$

$= R(2 \sin\frac{1}{2}A \cos\frac{1}{2}A + (\sin B + \sin C))$

$= R(2 \sin\frac{1}{2}A \cos\frac{1}{2}A + 2 \sin\frac{1}{2}(B + C) \cos\frac{1}{2}(B - C))$

$= 2R(\sin\frac{1}{2}A \cos\frac{1}{2}A + \cos\frac{1}{2}A \cos\frac{1}{2}(B - C))$

$= 2R \cos\frac{1}{2}A(\sin\frac{1}{2}A + \cos\frac{1}{2}(B - C))$

$= 2R \cos\frac{1}{2}A(\cos\frac{1}{2}(B + C) + \cos\frac{1}{2}(B - C))$

$= 2R \cos\frac{1}{2}A(2 \cos\frac{1}{2}B \cos\frac{1}{2}C)$

$= 4R \cos\frac{1}{2}A \cos\frac{1}{2}B \cos\frac{1}{2}C = $ RHS

Exercise 13.2 (page 137)

1 $\frac{3}{4}\sqrt{15}\,\text{cm}^2$, $\frac{1}{6}\sqrt{15}\,\text{cm}$, $\frac{3}{10}\sqrt{15}\,\text{cm}$, $\frac{1}{2}\sqrt{15}\,\text{cm}$, $\frac{3}{2}\sqrt{15}\,\text{cm}$

2 $\dfrac{6\sqrt{3}}{7+\sqrt{13}} = \dfrac{\sqrt{3}(7-\sqrt{13})}{6}\,\text{cm}$

3 Using Equations 3 and 4:

$$\text{LHS} = r r_A r_B r_C$$
$$= \frac{\Delta}{s} \times \frac{\Delta}{s-a} \times \frac{\Delta}{s-b} \times \frac{\Delta}{s-c}$$
$$= \frac{\Delta^4}{\Delta^2} = \Delta^2 = \text{RHS}$$

4 $$\text{LHS} = \frac{\Delta}{s}$$
$$= \frac{\frac{1}{2}ab\sin C}{\frac{1}{2}(a+b+c)}$$
$$= \frac{2R\sin A \times 2R\sin B \times \sin C}{2R\sin A + 2R\sin B + 2R\sin C}$$
$$= 2R\,\frac{\sin A \sin B \sin C}{\sin A + \sin B + \sin C}$$
$$= 2R\,\frac{2\sin\frac{1}{2}A\cos\frac{1}{2}A \times 2\sin\frac{1}{2}B\cos\frac{1}{2}B \times 2\sin\frac{1}{2}C\cos\frac{1}{2}C}{2\sin\frac{1}{2}(A+B)\cos\frac{1}{2}(A-B) + 2\sin\frac{1}{2}C\cos\frac{1}{2}C}$$
$$= 2R\,\frac{8\sin\frac{1}{2}A\cos\frac{1}{2}A\sin\frac{1}{2}B\cos\frac{1}{2}B\sin\frac{1}{2}C\cos\frac{1}{2}C}{2\cos\frac{1}{2}C\left(\cos\frac{1}{2}(A-B) + \sin\frac{1}{2}C\right)}$$
$$= 8R\,\frac{\sin\frac{1}{2}A\cos\frac{1}{2}A\sin\frac{1}{2}B\cos\frac{1}{2}B\sin\frac{1}{2}C}{\cos\frac{1}{2}(A-B) + \cos\frac{1}{2}(A+B)}$$
$$= 8R\,\frac{\sin\frac{1}{2}A\cos\frac{1}{2}A\sin\frac{1}{2}B\cos\frac{1}{2}B\sin\frac{1}{2}C}{2\cos\frac{1}{2}A\cos\frac{1}{2}B}$$
$$= 4R\sin\frac{1}{2}A\sin\frac{1}{2}B\sin\frac{1}{2}C = \text{RHS}$$

5 LHS $= \dfrac{\Delta}{s-c}$

$= \dfrac{\frac{1}{2}ab\sin C}{\frac{1}{2}(a+b-c)}$

$= \dfrac{2R\sin A \times 2R\sin B \times \sin C}{2R\sin A \times 2R\sin B - \sin C}$

$= 2R\,\dfrac{\sin A \sin B \sin C}{\sin A + \sin B - \sin C}$

$= 2R\,\dfrac{2\sin\frac{1}{2}A\cos\frac{1}{2}A \times 2\sin\frac{1}{2}B\cos\frac{1}{2}B \times 2\sin\frac{1}{2}C\cos\frac{1}{2}C}{2\sin\frac{1}{2}(A+B)\cos\frac{1}{2}(A-B) - 2\sin\frac{1}{2}C\cos\frac{1}{2}C}$

$= 2R\,\dfrac{8\sin\frac{1}{2}A\cos\frac{1}{2}A\sin\frac{1}{2}B\cos\frac{1}{2}B\sin\frac{1}{2}C\cos\frac{1}{2}C}{2\cos\frac{1}{2}C\big(\cos\frac{1}{2}(A-B) - \sin\frac{1}{2}C\big)}$

$= 8R\,\dfrac{\sin\frac{1}{2}A\cos\frac{1}{2}A\sin\frac{1}{2}B\cos\frac{1}{2}B\sin\frac{1}{2}C}{\cos\frac{1}{2}(A-B) - \cos\frac{1}{2}(A+B)}$

$= 8R\,\dfrac{\sin\frac{1}{2}A\cos\frac{1}{2}A\sin\frac{1}{2}B\cos\frac{1}{2}B\sin\frac{1}{2}C}{2\sin\frac{1}{2}A\sin\frac{1}{2}B}$

$= 4R\cos\frac{1}{2}A\cos\frac{1}{2}B\sin\frac{1}{2}C = $ RHS

6 $\cos\frac{1}{2}A = \sqrt{\dfrac{s(s-a)}{bc}}$; $\tan\frac{1}{2}A = \sqrt{\dfrac{(s-b)(s-c)}{s(s-a)}}$

7 RHS $= \dfrac{b-c}{b+c}\cot\frac{1}{2}A$

$= \dfrac{2R\sin B - 2R\sin C}{2R\sin B + 2R\sin C} \times \dfrac{\cos\frac{1}{2}A}{\sin\frac{1}{2}A}$

$= \dfrac{\sin B - \sin C}{\sin B + \sin C} \times \dfrac{\cos\frac{1}{2}A}{\sin\frac{1}{2}A}$

$= \dfrac{2\cos\frac{1}{2}(B+C)\sin\frac{1}{2}(B-C)}{2\sin\frac{1}{2}(B+C)\cos\frac{1}{2}(B-C)} \times \dfrac{\cos\frac{1}{2}A}{\sin\frac{1}{2}A}$

$= \dfrac{\sin\frac{1}{2}A\sin\frac{1}{2}(B-C)}{\cos\frac{1}{2}A\cos\frac{1}{2}(B-C)} \times \dfrac{\cos\frac{1}{2}A}{\sin\frac{1}{2}A}$

$= \dfrac{\sin\frac{1}{2}(B-C)}{\cos\frac{1}{2}(B-C)}$

$= \tan\frac{1}{2}(B-C) = $ LHS

8 RHS $= (s-a) \times \sqrt{\dfrac{(s-b)(s-c)}{s(s-a)}}$

$= \sqrt{\dfrac{(s-a)(s-b)(s-c)}{s}}$

$= \dfrac{1}{s} \sqrt{s(s-a)(s-b)(s-c)}$

$= \dfrac{\Delta}{s}$

$= r = \text{LHS}$

Exercise 14 (page 144)

In all the answers to this exercise, n and m are integers. It is possible that your answer might take a different form from the one given and still be correct.

1 $180n + (-1)^n 60$	**2** $360n \pm 60$
3 $180n - 45$	**4** $180m$ or $180m + 120$
5 $120n + 60$	**6** $60n - 30$
7 $180n$	**8** $72n$
9 $180m$ or $90m + 45$	**10** $(4m+1)45$ or $(4m+1)22\frac{1}{2}$
11 $\frac{1}{2}n\pi - \frac{1}{12}(-1)^n\pi$	**12** $\frac{2}{3}n\pi \pm \frac{1}{6}\pi$
13 $\frac{1}{2}n\pi - \frac{1}{4}\pi$	**14** $\frac{2}{3}m\pi$ or $\frac{1}{3}(2m+1)\pi - \frac{1}{9}\pi$
15 $\frac{1}{2}n\pi$	**16** $\frac{1}{2}n\pi + \frac{3}{8}\pi$
17 $\frac{1}{2}n\pi + \frac{1}{4}\pi$	**18** $\frac{1}{10}(4n+1)\pi$ or $\frac{1}{2}(4n-1)\pi$
19 $\frac{1}{2}n\pi$	**20** $\frac{2}{3}n\pi$

teach yourself ®

the A-Z of teach yourself

How to Win at Poker
HTML Publishing on the WWW
Human Anatomy & Physiology
Hungarian
Icelandic
Indian Head Massage
Indonesian
Information Technology, 101 Key Ideas
Internet, The
Irish
Islam
Italian
Italian, Beginner's
Italian Grammar
Italian, Instant
Italian Grammar, Quick Fix
Italian Language, Life & Culture
Italian Verbs
Italian Vocabulary
Japanese
Japanese, Beginner's
Japanese, Instant
Japanese Language, Life & Culture
Japanese Script, Beginner's
Java
Jewellery Making
Judaism
Korean
Latin
Latin American Spanish
Latin, Beginner's
Latin Dictionary
Latin Grammar
Letter Writing Skills
Linguistics
Linguistics, 101 Key Ideas
Literature, 101 Key Ideas
Mah Jong
Managing Stress
Marketing
Massage
Mathematics
Mathematics, Basic
Media Studies
Meditation
Mosaics
Music Theory
Needlecraft
Negotiating
Nepali
Norwegian
Origami
Panjabi
Persian, Modern

Philosophy
Philosophy of Mind
Philosophy of Religion
Philosophy of Science
Philosophy, 101 Key Ideas
Photography
Photoshop
Physics
Piano
Planets
Planning Your Wedding
Polish
Politics
Portuguese
Portuguese, Beginner's
Portuguese Grammar
Portuguese, Brazilian
Portuguese, Instant
Portuguese Language Life & Culture
Postmodernism
Pottery
Powerpoint 2002
Presenting for Professionals
Project Management
Psychology
Psychology, 101 Key Ideas
Psychology, Applied
Quark Xpress
Quilting
Recruitment
Reflexology
Reiki
Relaxation
Retaining Staff
Romanian
Romanian BK/CASS PK 2ED
Romanian CASS 2ED
Russian
Russian, Beginner's
Russian Grammar
Russian, Instant
Russian New Edition BOOK
Russian Language Life & Culture
Russian Script, Beginner's
Sanskrit
Screenwriting
Serbian
Setting up a Small Business
Shorthand, Pitman 2000
Sikhism
Spanish
Spanish, Beginner's
Spanish Grammar
Spanish Grammar, Quick Fix

Spanish, Instant
Spanish Language, Life & Culture
Spanish Starter Kit
Spanish Verbs
Spanish Vocabulary
Speaking on Special Occasions
Speed Reading
Statistical Research
Statistics
Swahili
Swahili Dictionary
Swedish
Tagalog
Tai Chi
Tantric Sex
Teaching English as a Foreign
   Language
Teams and Team-Working
Thai
Time Management
Tracing your Family History
Travel Writing
Trigonometry
Turkish
Turkish Beginner's
Typing
Ukrainian
Urdu
Urdu Script, Beginner's
Vietnamese
Volcanoes
Watercolour Painting
Weight Control through Diet and
   Exercise
Welsh
Welsh Dictionary
Welsh Language Life & Culture
Wills and Probate
Wine Tasting
Winning at Job Interviews
Word 2002
World Faiths
Writing a Novel
Writing for Children
Writing Poetry
Xhosa
Yoga
Zen
Zulu

available from bookshops and on-line retailers